insight into how these theories could be fundamentally linked through space-time geometry."

Philip Turner, Director, Centre for Plant Science and Biopolymer Research, Edinburgh Napier University

"Laurent Nottale proposes that we look at the concept of fractals to make relativity, extended further yet, the fundamental principle on which to base quantum mechanics. After the relativity of time and space, he has tackled the relativity of scale, putting into question much of what we thought we knew."

Pierre Bonnaure, *Futuribles*

"Developments in geometry have often enabled progress in physics, especially when concerning relativity. Non-Euclidean geometry, geometrical systems where the plane is a sphere, made it possible for Einstein to devise his theory of curved space. Today, a new geometry, fractal geometry, allows us to propose a theory of fractal space."

Idées clés, by Business Digest

Laurent Nottale
The Relativity of All Things

Laurent Nottale is a former Director of Research at the French National Center for Scientific Research and a visiting Researcher at the Paris Observatory. His technical works which are available in English include *Scale Relativity and Fractal Space-Time* and *Fractal Space-Time and Microphysics*.

THE RELATIVITY OF ALL THINGS

THE RELATIVITY
OF ALL THINGS

BEYOND SPACETIME

LAURENT NOTTALE

Translated by Mia Chen
with Afterword by Charles Alunni

PERSISTENT PRESS

est. 2012

CONTENTS

PREFACE   xiii

PART ONE: THEORIES OF RELATIVITY OF MOTION

Chapter 1: The Precursor: 3
          Copernicus and Heliocentrism

Chapter 2: The Discoverer: 8
          Galileo and the Laws of Inertia

Chapter 3: Newton: 13
          Universal Gravitation
          and Absolute Space

    Universal Gravitation   13
    Differential Calculus   16
    The Absolute Space of Newton   20

Chapter 4: Special Relativity: 22

    Poincaré and Relativity   26
    Einstein and Special Relativity   36

Chapter 5: Einstein's General Relativity: 46

    The Equivalence Principle   47
    The Problem of Inertial Forces   49
    Relativity of Gravity   51
    Generalizing the Principle of Relativity 53
    Relativity of Geometry   54
    Curved Spacetime   55
    Geodesics   58
    The Sources of Gravity   60
    Spacetime Equations   61

PART TWO: THE PRINCIPLE OF RELATIVITY

Chapter 6: What the Principle of Relativity Means   67

    The Laws of Nature   68
    Toward More and More Fundamental Laws   70
    Relativity as Unifying Principle   71
    Reference Systems   74
    Relativity of the Frame of References State   76
    Relativity, Not Relativism   78
    Relativity and Construction of Laws   79
    From the Relative to the Absolute   81

Chapter 7: Toward a New Extension of Relativity   82

    Relativity as Method and as Way of Thinking   82
    Beyond Spacetime   85

PART THREE: QUANTUM MECHANICS

Chapter 8: Principal Axioms   91
           of Quantum Mechanics

    Probability and the Wave Function   93
    The Correspondence Principle   99
    Schrödinger's Equation   100
    Collapse of the Wave Function   100

Chapter 9: The Paradox of Quantum Properties   102

    Einstein-de Broglie Relations   102
    Heisenberg Relations   103
    Spin   104
    Indistinguishability of Identical Particles   106
    Inseparability, Entanglement, and Nonlocality   107

Chapter 10: Einstein and Quantum Theory   110

**PART FOUR: RELATIVITY, FRACTAL SPACETIME,
AND QUANTUM MECHANICS**

Chapter 11: Scales in Nature   119

Chapter 12: The Concepts of Point and Instant   133
　　　　　in Physics

Chapter 13: Fractal Geometry   137

Topological Dimension and Fractal Dimension   137
Where Do Fractals Come From?   148
Optimization Under Constraint   148
Renormalization Group   151
Dynamical Chaos   152
Beyond Differential Geometry   156
From Fractal Objects to Fractal Spaces   159

Chapter 14: The "Missing Link"   161
　　　　　in Quantum Theory

The Nature of Quantum Spacetime   162
Toward a Nondifferentiable Spacetime   164
Feynman and the Return to a Spatiotemporal   166
　　Representation
An Extension of Relativity?   168

Chapter 15: Scale Relativity   169

The Structure of the Electron   170
Resolutions in Physics   173
Relativity of Scales   174
Changes of Units   174
Change of Resolution   177
Universality of the Heisenberg Relation   178
The Principle of Scale Relativity   181

Chapter 16: Special Scale Relativity   184

"Galilean" Scale Laws   185
"Lorentzian" Scale Laws   187
The Planck Length as Invariant, Unsurpassable,   189
    and Unreachable Scale

Chapter 17: Scale Relativity   195
                  and Quantum Theory

Introduction   195
Fractal Geodesics   196
Infinite Geodesics and Probabilities   196
The Nature of Particles   199
Quantum/Classical Duality   202
Irreversibility and the Complex Character   203
    of the Wave Function
Free Motion and the Schrödinger Equation   206

PART FIVE: FROM ELEMENTARY PARTICLES
                TO THE LARGE STRUCTURES
                OF THE UNIVERSE

Chapter 18: Particle Physics   216

Grand Unification   217
Charge Variation in Function of Scale   219
Nature of the Electric Charge and the   221
    Electromagnetic Field

Chapter 19: Cosmology   225

The Problem of the Origin   226
Horizon and Causality   227
Maximal Scale and Cosmological Constant   229
Energy Density of the Quantum Vacuum   231

Chapter 20: Formation and Evolution   237
          of Structures

    Gravitational Structures   238
    The Kepler Problem   239
    The Solar System   240
    Extrasolar Planets   245
    Planets Around a Pulsar   248
    Structure Formation   249
    Planetary Nebulae   251
    Galaxies   252
    Other Sciences   255

CONCLUSION   267

AFTERWORD by Charles Alunni   275
    Theories of Relativity: What It Means
      for Philosophy

NOTES   302

BIBLIOGRAPHY   338

INDEX   346

List of Figures

Figure 1 Geodesic on the surface of a sphere   58
Figure 2 Probability amplitude   94
Figure 3 "Scale of scales" in nature   121
Figure 4 Fractal curve   130
Figure 5 "Peano Curve"   138
Figure 6 Similarity of non-fractal objects   139
Figure 7 "Koch curve"   140
Figure 8 Fractal curve of dimension 1.5   141
Figure 9 Fractal function   142
Figure 10 Zooming in on two fractal curves   143
Figure 11 Fractal surface   144
Figure 12 Fractal/Non-fractal transition   145
Figure 13 Fractal curve in space   146
Figure 14 Fractal curve of variable dimension   147
Figure 15 Model of self-similar growth   150
Figure 16 Lorenz's chaotic attractor   154
Figure 17 Fluctuations of Jupiter's force   155
Figure 18 Virtual particles inside electron   171
Figure 19 Scale dependence of electric charge   172
Figure 20 Fractal laws in special scale relativity   190
Figure 21 New dilation laws   191
Figure 22 Geodesics on a fractal surface   198
Figure 23 Time evolution of a fractal coordinate   202
Figure 24 Length vs. energy in scale relativity   217
Figure 25 Variation of charges in scale relativity   218
Figure 26 Mass-charge relation for the electron   223
Figure 27 Light cones in special scale relativity   228
Figure 28 Scale dependence of fractal dimension   231
Figure 29 Scale dependence of vacuum energy   236
Figure 30 Observations vs. theory in solar system   241
Figure 31 Probability peaks in Kuiper Belt   244
Figure 32 Distribution of first exoplanets (1996)   246
Figure 33 Distribution of exoplanets (2008 data)   247

Figure 34 Exoplanet velocities (2018 data)   247
Figure 35 System of three pulsar planets   248
Figure 36 Quantum harmonic oscillator   250
Figure 37 Quantized shapes of planetary nebulae   252
Figure 38 Rotation velocities of spiral galaxies   253
Figure 39 Velocity differences in galaxy pairs   254
Figure 40 Variable fractal dimension in geography   256
Figure 41 Acceleration of arctic sea ice melting   258
Figure 42 Opening of a quantum flower   259
Figure 43 Spontaneous cell-like division   260
Figure 44 Log-periodicity on tree of life   263
Figure 45 Main steps of embryogenesis   264
Figure 46 Log-periodicity of civilizations   265
Figure 47 Log-periodicity of seismic aftershocks   266

PREFACE

What is relativity? The word evokes thoughts of Einstein, $E = mc^2$, the twin paradox. Behind these commonplaces lies one of the most marvelous adventures of science and the mind. It marks a shift in worldview which can be traced back to Galileo, and crowns two thousand years of human endeavor to remove the veil of illusion. Four centuries ago, Galileo discovered that 'motion is like nothing.' In other words, motion or rest have no proper existence; only the motion of one body relative to another can have meaning. What ultimately matters is the relationship between two objects, not their absolute properties, which are emptied of physical significance.

This book first follows the idea of relativity through history, going from Copernicus to Einstein, encountering along the way such great names as Galileo, Descartes, Huygens, Mach, and Poincaré. One discovers that relativity, a philosophical and scientific postulate, is a unifying principle as well, a method of constructing the laws of physics, a mode of diagnosing its crises, even a way of thought.

In 1916, Einstein defined the principle of relativity in this way: "The laws of nature should be valid in all systems of reference, under all conditions." This founding principle of physics, bearing upon even the existence of fundamental laws, must play a leading role in the quest for meaning and the search for a true understanding of natural phenomena.

With the triumph of Einstein's theory of general relativity, which encompasses all of classical mechanics and makes gravitation the most precise theory in today's physics, has the theory of relativity said its last word? The development of physics over the last century might lead us to believe so.

In fact, we will recall how the domain of the minuscule, the scale of atoms, their nuclei, and elementary particles, has only been able to be described in terms of a completely new framework, that of quantum mechanics. Yet it would seem impossible to deduce its postulates, its methods, even its way of reasoning from the concepts of classical theory, that is to say, from the principle of relativity of motion.

Quantum mechanics remains essentially axiomatic in theory, and cannot be, even today, considered fully understood; its numerous postulates seem arbitrary in nature, when one would prefer to see them brought to a first principle.

The idea developed in the second half of this book suggests that there does indeed exist a fundamental principle on which to base quantum mechanics: the principle of relativity itself. But this is only possible by generalizing further, by applying relativity not only to displacement and motion, but also to changes of scale: that is to say, to the transformations which lead us from the scale of the very small to the very large, or the inverse. Examples of such transformations can be found in microphysics in the use of a loupe, a microscope, or a particle accelerator, and in macroscopic physics, in the use of glasses or a telescope.

What can we expect from such an extension of our frame of mind? It is not a matter of putting into question the discoveries of physics. On the contrary: it is only by basing ourselves upon them that we are able to go further. In such a way, relativity of motion has passed from the flat and absolute space of Newton to the relativistic, curved spacetime of Einstein. Similarly, I will demonstrate how the concept of "scale relativity" leads to a new spatiotemporal geometry: spacetime becomes fractal, which is to say that it possesses structures at every scale.

The book will conclude with some of the numerous consequences which follow from this new approach, going from the possibility of a renewed understanding of quantum mechanics and of its generalization toward very high energies, to the proposition of original physical effects, in particular in the domain of astrophysics. Indeed, one of the fundamental implications of scale relativity posits that certain aspects of quantum theory could be equally applied to macroscopic scales, but with a different interpretation.

We will thus see how this theory, considered as a description of the tendency for a system to form structures, from planetary systems up to the great formations of the Universe, predicts many hidden effects, some of which have now been revealed by astronomic observations.

I would like to warmly thank Pierre Grou, Jean Chaline, Gérard Schumacher, and Georges Alecian, my friends and colleagues, who kindly read early (French) versions of the present work and offered their critiques and suggestions. The exposition has been greatly improved as a result.

Neither this nor my first book (*L'Univers et la lumière* [Light and the Universe], Flammarion, 1995) would have been possible without the enthusiasm that Louis Audibert brought to them, without his patient reading, his advice, and his stimulating observations. I am very grateful to him.

Regarding the new English version of this book, I am immensely grateful to the Editor-in-Chief of Persistent Press, Grant Maxwell, for his kind involvement in its publication, to Mia Chen for her excellent translation, and to Charles Alunni, for his unfailing friendship, his continuous interest in the scale relativity theory and his deep philosophical analyses, and for having accepted to write an afterword to this new edition.

This book is dedicated to all those to whom I am connected.

Part One
Theories of
Relativity of Motion

Chapter 1
The Precursor:
Copernicus and Heliocentrism

In some regard, relativity begins with Copernicus. To be sure, there is no explicit statement of relativity in his work: relativity's discovery without a doubt belongs to Galileo. But it is no accident that the trial of Galileo focused on his defense of Copernican heliocentrism: what was important for Galileo was the question of the Earth's motion. (*"And yet it moves!"*)

If Copernicus is indeed a precursor to relativity, it is because decentering the Earth is the first step necessary for comprehending the relativity of coordinate systems.

Before Copernicus, the world had a center. According to one of the dominant ideas of pre-Copernican physics (physics termed, perhaps unfairly and over-simplistically, Aristotelian), the natural state of terrestrial bodies, the state toward which they all tend, is rest. All bodies have a tendency to approach the center of the universe and to cease all motion. Such a theory was already, in its time, unifying: it explained why the Earth's center coincided with the center of the universe, which was the center of all celestial motion (the latter being imperishable, not subject to terrestrial "corruption"). It thus brought together into a single description the various phenomena which we currently attribute either to gravitation (the falling of bodies) or to dissipation (the friction which masks the laws of inertia).

Copernicus, in decentering the Earth, began the process which would shatter this entire construction. He put the Sun at the center of the universe. A few decades later, Kepler discovered that the planets moved not in circles, but ellipses, and that the Sun was not at the center of their orbits, but at one of their focal points. Galileo discovered the laws of inertia; Descartes stated them formally, and applied them to celestial bodies, affirming the law of conservation of momentum. The Sun itself, accordingly, could not remain immobile at the focal point, but must also move in a small ellipse. Little by little, it became apparent that the Earth was not the center of the universe, then that the Sun was not either, and finally, that the center of the universe, in reality, did not exist. In post-Copernican physics, there is no longer any particular point of origin to which to compare the position of other objects. Every point can equally serve as the origin of a coordinate system. The choice of this origin is completely arbitrary, and any choice is possible and equivalent from the point of view of the fundamental laws of physics.

It is in this sense that relativity starts with Copernicus, even if in an implicit fashion, and only if later developments would make it manifest: the Copernican revolution implies the relativity of position. No position is absolute; the same laws are applicable no matter what the origin of the coordinate system is. There is thus an invariance of the laws when shifting this system of coordinates, what in mathematical terms are called *translations*.

But the Copernican revolution goes even further, toppling the entire edifice of received wisdom. The Earth is brought down to the level of a planet like any other, which implies that it moves, opening the way toward Galilean relativity. The Aristotelian division of terrestrial and celestial laws was no longer valid. Before Copernicus, the heavens existed only as "object." They took the shape of concentric spheres, composed of a special substance, crystalline, indestructible, perfectly hard and transparent. Even if Copernicus himself did not explicitly cast doubt upon this belief, the consequences of his new system on the nature of the heavens quickly became apparent. Tycho Brahe, in 1588, declared that "the apparatus of the heavens is not a hard and impenetrable substance filled with real spheres, as has been the belief of most people until now," but that it was something "fluid and simple" through which the planets could freely move. The two novae of 1604 and 1672 brought fresh evidence of the heavens being themselves subject to corruption and change.

When the idea of a center of the universe disappeared, questions about its outer limits quickly arose as well. Thomas Digges, in 1576, disseminated Copernican theory to a wide audience in England, while drawing his own conclusions and abandoning the idea of the sphere of fixed objects: for him, already, the Universe was infinite, and the stars distant suns. Then, Giordano Bruno would make the same arguments, and added his own, claiming that life existed everywhere throughout the Universe. These ideas, along with his denial of the divinity of Christ, earned him the stake.

We know today that there exists a profound connection between relativity of positions (and of instants when applied to time) and the existence of conservation laws. This connection could even be said to constitute the essence of modern physics, which can be traced back to the Copernican revolution, principally through the work of Descartes. His physico-mathematical system clearly upheld Copernican theory with a coherence that would contribute greatly to its definitive acceptance in Europe. In his system, it would be impossible to set a limit to the extent of the Universe, making it infinite. Similarly, it would be impossible to set a limit to the division into the infinitely small (but Descartes applied this to matter and not space, which led him to reject the notion of atoms as well as of a vacuum). We will return to these questions when discussing the antinomies of Kant and then the foundations of scale relativity. What is important for us here is that Descartes, still motivated to give his vision of the universe coherence, was led to state: "God always preserves the same amount of mass and momentum."

The concepts of energy and momentum thus took their modern form in terms of magnitudes keeping constant over the course of time. It was an essential development for the construction of the completely new science that Newton would synthesize forty years later. With this "new world system," one would seemingly be forced to abandon the idea of eternity and imperishability within celestial laws, even those governing the heavens, in favor of the variability, instability, and "corruption" characteristic of terrestrial laws. What Descartes demonstrated, with his discovery of the conservation of matter and of motion (he was also the first to state the principle of inertia in its present form of conservation of the speed and direction of motion for a free body, Galileo only allowing for circular motion to continue perpetually), was that, beyond the appearance of change, the deeper structures stayed the same. Temporal evolution, of which the essence was change, nevertheless allowed for and indeed necessarily accompanied the inalterability of some quantities, magnitudes which would stay invariant despite seeming to change.

We now know that these invariants, the most fundamental being energy, linear momentum, and angular momentum, are a direct consequence of the relativity of instants, positions, and orientations. Because any instant whatsoever can be taken for the temporal origin of a physical experiment, and the experiment produces the same result now, a year ago, or an hour from now, there must be conservation of energy. Energy is born from the uniformity of time, that is to say, from the relativity of the temporal origin of coordinate systems.

Similarly, the spatial origin of a coordinate system according to which an experiment is measured can take place here, ten meters to the right, a kilometer in front, or five centimeters below, and nothing will change in its results. In this case, it is not motion along a single axis, and thus a single quantity conserved (as is the case for time and energy), but translations along three possible spatial axes. There must then be three magnitudes, the three components of linear momentum (the product of mass and speed), which arise from the uniformity of space, that is, the relativity of positions. Each one in itself is not invariant, for its numerical value depends on the choice of orientation of the coordinate system; only when taken together, forming a vector quantity, do they become an invariant.

Finally, all orientations of the axes of a frame of reference are equivalent, which implies the appearance of three new magnitudes, the components of angular momentum. The conservation of angular momentum (well known from the example of a figure skater lowering his or her arms in order to spin faster) is thus tied to the isotropy of space, that is to say, to the relativity of orientation.

At the beginning of the twentieth century, Emmy Noether formalized this connection between symmetries and conservation laws in what constitutes one of the most fundamental theorems of theoretical physics. It can in fact be generalized to symmetries other than those of time and space, and thus represents one of the key principles in the search for invariants in physics.

Chapter 2
The Discoverer:
Galileo and the Laws of Inertia

Even though the word "relativity" has become indissociable from Einstein's name, and in the other direction, Copernicus can be considered the theory's precursor, its history truly begins with Galileo. Relativity of position may be a consequence of Copernicus's discovery, but he neither observed nor described it directly. Relativity of motion, however, was explicitly discovered, described, and studied by Galileo, and he well understood its significance.

It is not by chance that relativity of motion, as we currently understand it, was discovered by Galileo, the most steadfast proponent of Copernican theory. Relativistic thinking can simply not occur if there is universal belief (often implicit) in an absolute frame of reference. Galileo explicitly makes this connection, starting his *Dialogue* by directly stating the Earth's movement:

> Copernicus places the earth among the movable heavenly bodies, making it a globe like a planet.[1]

Of course, the relativity of motion had its precursors, going from Copernicus all the way back to Aristotle, who observed that a ship on a river is "as if in a vase," caught up in the motion of the river. But Aristotle only wanted to emphasize the difficulty involved in discovering true motion, whose existence he did not doubt. William of

Conches, in the twelfth century, already observed that if we move without having a motionless object with which to compare our position, we do not feel ourselves moving. Here again, there is no doubt of the meaning of "motionlessness," only an insistence on the fact that motion must be seen, as it seems to elude all the other senses. It would be the same for Nicolas Oresme, who pointed out that the Earth could be moving without our realizing it: we can distinguish its motion by sight, but without knowing which is moving, the one who observes (the Earth) or the objects observed (the stars). For him, though, the goal is to highlight the difficulty involved in demonstrating the immobility of the Earth rather than to prove its movement.

Finally, for Copernicus, who returned to the argument of the boat, this relativism allowed him to construct his new cosmology without objectors being able to consider the posited motion of the Earth as a reason to contradict it (due to its apparent stasis in the old cosmology and everyday experience):

> Why therefore do we still hesitate to concede movement to [it]? . . . And why should we not admit that the daily revolution itself is apparent in the heaven [sic], but real in the Earth; and the case is just as if Virgil's Aeneas were saying 'We sail out from the harbour, and the land and cities recede'? For when a ship is floating along in calm weather, everything which is outside her is perceived by those who are sailing as moving by a reflection of that motion, and on the other hand they think that they are at rest along with everything that is with them.[2]

But Galileo goes much further than his predecessors in posing a thesis so revolutionary that even today it can seem shocking to those who come across it for the first time. In this demonstration of the relativity of motion, it is not a matter of showing the difficulty of detecting "true motion" or "true rest," but of something much more essential. What Galileo affirms is that "true motion" does not exist; the nature of motion can only be relative. No local experiment can determine the motion or stasis of a body, so that only the inertial movement of a body *in relation to another* has meaning:

> Motion, in so far as it is and acts as motion, to that extent exists relatively to things that lack it; and among things which all share equally in any motion, it does not act, and is as if it did not exist.[3]

Galileo, particularly in the *Dialogue*, provides a multitude of examples and thought experiments to convince his readers of this astonishing fact. No longer is it a case of a simple undetectability of motion to our senses, but a veritable impossibility, physically speaking, to detect it, whatever the type of experiment devised. On this matter, the example of a ship's hold is striking:

> Shut yourself up with some friend in the main cabin below decks on some large ship, and have with you there some flies, butterflies, and other small flying animals. Have a large bowl of water with some fish in it; hang up a bottle that empties drop by drop into a wide vessel beneath it. With the ship standing still, observe carefully how the little animals fly with equal speed to all sides of the cabin. The fish swim indifferently in all directions; the drops fall into the vessel beneath; and, in throwing something to your friend, you need throw it no more strongly in one direction than another, the distances being equal; jumping with your feet together, you pass equal spaces in every direction. When you have observed all these things carefully (though there is no doubt that when the ship is standing still everything must happen in this way), have the ship proceed with any speed you like, so long as the motion is uniform and not fluctuating this way and that. You will discover not the least change in all the effects named, nor could you tell from any of them whether the ship was moving or standing still . . . The cause of all these correspondences of

effects is the fact that the ship's motion is common to all the things contained in it, and to the air also.[4]

Galileo gives another remarkable and particularly impressive example, one which still merits a place today in the instruction of physics:

SAGREDO : There has just occurred to me a certain fantasy which passed through my imagination one day . . . Perhaps it may be of some help in explaining how this motion in common is nonoperative and remains as if nonexistent to everything that participates in it . . . If the point of a pen had been on the ship during my whole voyage from Venice to Alexandretta and had had the property of leaving visible marks of its whole trip, what trace—what mark—what line would it have left?

SIMPLICIO: It would have left a line extending from Venice to there . . . Without an error of any moment it could be called part of a perfect arc.

SAGREDO: . . . Now if I had had that same pen continually in my hand, and had moved it only a little sometimes this way or that, what alteration should I have brought into the main extent of this line?

SIMPLICIO: Less than that which would be given to a straight line a thousand yards long which deviated from absolute straightness here and there by a flea's eye.

SAGREDO: Then if an artist had begun drawing with that pen on a sheet of paper when we left the port and had continued doing so all the way to Alexandretta, he would have been able to derive from the pen's motion a whole narrative of many figures, completely traced and sketched in thousands of directions, with landscapes, buildings, animals, and other things. Yet the actual, real, essential movement marked by the pen point would have been only a line; long, indeed, but very simple. But as to the artist's own actions, these would have been conducted exactly the same as if the ship had been standing still. The reason that of the pen's long motion no trace would remain except the marks drawn upon the paper is that the gross motion from Venice to Alexandretta was common to the paper, the pen, and everything else in the ship. But the small motions back and forth, to right and left, communicated by the artist's fingers to the pen but not to the paper, and belonging to the former alone, could thereby leave a trace on the paper which remained stationary to those motions.[5]

The conclusion is clear. Neither of the two points of view can be considered more "true" than the other. Had a great sheet of paper been laid between Venice and Alexandretta, and had the painter drawn on carbon paper so as to be able to mark at the same time the sheet attached to him and the ship, and the great immobile sheet with respect to the shores, the final result would have been *two* traces of movement, the long straight line running over thousands of kilometers and the picture with all its figures. Neither of the two reference points could be considered to be moving by themselves; only with respect to the other can each be considered to be moving.

Why, then, did this illusion persist? Why did humanity remain so long in error on a subject so fundamental? The response is without a doubt the unique and limited character of the frames of reference within which we ordinarily live. The existence of the terrestrial ground, and the simple fact that it is much larger than us, makes us grant this reference point a special role. Now, as before, common language can bypass the subtleties of relativity: most of the time, it is possible to state that one is moving or that one is immobile without having to say with respect to what simply because it is obvious that it is with respect to the Earth.

Chapter 3
Newton:
Universal Gravitation and
Absolute Space

The theory of universal gravitation without a doubt constitutes one of the most impressive unifications in the history of physics. It is not at all intuitive to understand that phenomena as apparently contradictory as the systematic falling down of bodies on our Earth and the permanence of the Moon in the sky could be deduced from a single theory, and brought together in a single cause.

Even if the existence of a force of attraction between the Sun and the planets had been envisioned before Newton, it could well have been a new force, operating only on celestial bodies, or the force of magnetism, as Kepler proposed. It was hardly obvious that this force was precisely the same as that which makes all bodies on Earth fall down.

Universal Gravitation

Newton's logic can be summed up as follows. The apple appears to fall, but the Moon does not. How could the same universal law govern two opposite behaviors? Newton's answer is that in actuality *the Moon falls as well!* To be convinced of this, one must recall Galileo's laws of inertia (as they were reformulated by Descartes). If the Moon were to follow a free trajectory, its motion would be linear and uniform. It would then get further and further away from the Earth, instead of maintaining a constant distance. Thus, over any period of time, no matter how small, the Moon "falls" by the amount which separates a circle from its tangent. This logic allows the motion of the Moon to be brought together with that of the apple, and vice versa. From the point of view of relativistic theory, one can say that the difference observed between different phenomena (in this case, they even appear to be opposite) is not the manifestation of different laws, but, on the contrary, of a single law which is applied in different *conditions* (meaning within coordinate systems of which the "state" is different).

Regarding this law, Newton postulated that between any two bodies there existed a force of attraction which varies directly with their masses and with the inverse square of the distance between them. The constant of proportionality governing this relation is the universal constant of gravitation G. This hypothesis enables the proof of Kepler's three laws governing the motion of planets, as well as developing novel predictions concerning celestial mechanics.

In his work, Newton was guided by the great physical principles established by Galileo and Descartes, for which he gave a more definitive mathematical formulation.[6] These principles were the law of inertia, the addition of superimposed velocities, the idea that force is not motion, but the change of motion, and finally the law of action and reaction.

Newton's work, under the form which Laplace gave to it, has become an archetype of field theory, which served as the model in the nineteenth century for the development of theories of electricity and magnetism.

Nevertheless, despite its success, Newton's formulation of gravitational theory was not without serious conceptual problems. The most obvious of these was the supposed *instantaneity* of gravity's action at a distance. Einstein's theory of special relativity would firmly establish the physical impossibility of such a transmission of force at an infinite speed, and with his theory of general relativity, he would solve the problem by showing that gravity propagates at the speed of light (although Poincaré had also obtained this result as early as 1900).

Furthermore, the foundations of Newton's theory lacked clarity. Why would the force depend on the inverse square of the distance and not some other relation? Why would it depend on the masses involved, and not other physical characteristics (chemical composition, shape, etc.)? Here again, general relativity would answer these questions of "why," while Newtonian gravitational theory seemed only to address the question of "how."

"I do not pretend to hypothesize," Newton admitted, to show how well aware he was of the problems and the incompleteness of his theory, while refusing to speculate on the origin of this action at a distance. He wanted an authentic explanation or nothing.

The problems of why and how are often, in fact, akin to an unending sequence of Russian dolls, each one containing a smaller one (we will return to this image later). What only seems to address the question of "how" (that is, a description closer in nature to a model than to a theory which rests on a first principle) can be addressed within a larger framework. In this way, Kepler's laws were first discovered in empirical fashion. However, once they were recognized, one could ask *why* the planets follow elliptical and not other trajectories, *why* the law of areas applies, *why* the square of a planet's orbital period is proportional to the cube of its semimajor axis. Newton's theory successfully answered these questions of why, for it justified these postulates using mathematical proof. But then, his explanation was based on deeper postulates, one of which was action of a force at a distance. Further questions emerged: why is the force proportional to the inverse square of the distance, and why does it depend upon mass alone and not other physical properties? Einstein's theory would respond to these questions, providing a mathematical proof of these laws. Yet general relativity itself contains its own lacunae, for example at the level of the source of gravity. The theory describes how masses bend spacetime, but does not explain why they do, which remains an unresolved problem.[7]

Differential Calculus

We have now arrived at one of the key moments in the history of science, and an essential development for the present work: the invention by Leibniz and Newton of differential calculus, which marks the birth of modern mathematical physics. What differential calculus brings to the description of the physical world is underpinned by an implicit hypothesis: that physical quantities, the most fundamental being positions and instants (in other words, spacetime), are differentiable. This means that the trajectory of a physical object in space possesses a tangent (or "slope")—in other words, that one can define a speed (the derivative of its position with respect to time). It is this hypothesis which the theory of scale relativity seeks to leave behind. However, before explaining what this abandonment entails, we must understand its deep physical significance and its fundamental role in the invention of mathematical physics.

The remarkable character of Newtonian theory can be found in the title of his magnum opus: *Philosophiae Naturalis Principia Mathematica* (Mathematical Principles of Natural Philosophy). The possibility of there being a science of physics in the sense which we understand it today was by no means self-evident at the time. It is true that Galileo had already stated in *The Assayer*:

> The book of Nature is . . . written in the language of mathematics.

But between simply describing and actually predicting, there is an enormous step to be made, which Leibniz and Newton's invention of differential calculus enabled. Newtonian theory did more than explain observations already made. Through calculation, it allowed the prediction of phenomena which only future observations would be able to verify, such as the return of Halley's comet and the discovery of Neptune by Le Verrier and Adams. Newton showed in his work that what seemed to fall solely under philosophical discourse (even when applied to the description of Nature) could be written in mathematical language. He did better yet when not limiting himself to already known mathematics, but instead constructing his own analytical tools when the existing ones were not enough.

The example of Newton shows that the question that Einstein and so many other physicists asked themselves—how is it that mathematics can be used to describe the world—is perhaps the wrong question. Only the existence of fundamental laws poses a problem (we will discuss, in the second part of this book, the difficulty of defining what a world without laws would be like). Once that is admitted, their expression naturally takes a form "compacted" into mathematical language.

One might consider it remarkable that, at times, the tools which a physicist needs to put his or her theory to work should already have been developed by mathematicians. A typical example is the concept of curved space, already developed in mathematics almost a century before Einstein used it to construct his theory of general relativity. This is to forget, however, that its inventor was Gauss himself, as great a physicist as mathematician, and that Riemann, who developed this theory, had envisioned that the curvature of space could bring about a new model of gravitation (but he lacked the necessary concept of space*time*, without which every attempt of this kind was destined for failure).

However, for every example of this kind, how many examples are there to the contrary—of Newton inventing differential calculus to unify celestial motion and terrestrial attraction, of Heisenberg (re-)inventing matrix calculus and thus founding quantum mechanics (it was only afterward that Born and Jordan informed him that this type of calculus already existed), of Einstein using the Wiener process to describe Brownian motion and demonstrate the limits of thermodynamics. In physics, mathematics is, for the most part, a tool which is used to put into place concepts and ideas which are not mathematical in their essence (although there certainly does exist an independent mathematical science which cannot be reduced to what the physicist can make use of).

Let us return to differential calculus. The first known attempt at infinitesimal calculus was Eudoxus of Cnidus's "method of exhaustion." Eudoxus, born around 408 B.C.E., was perhaps the greatest man of science in antiquity. We owe to him the theory of homocentric spheres which dominated astronomy for almost two thousand years by its ability to explain various celestial motions; the formalization of "Euclidean" geometry into "axioms" (posited) and "theorems" (deduced from axioms); the proof that irrational numbers cannot be expressed in the form of simple proportions. With the goal of calculating the volumes of complex solid bodies, such as a cone or sphere, Eudoxus divided them into infinitesimal sections, and then came up with a method of adding them back together, thus anticipating modern integral calculus. Kepler invented a similar method to calculate the volume of casks in his wine cellar and published a short treatise on the subject. Throughout the seventeenth century, mathematical research pursued the same subject. Descartes discovered that one could define curves in the form of an algebraic equation with the help of a grid on a plane. Fermat yet more closely approached infinitesimal calculus, and in particular solved the problems of minima and maxima.

Newton (using what he called the "calculus of fluxions") and Leibniz took very different approaches, which suffices in itself to dispel their mutual accusations of plagiarism. Newton sought to solve the problem of the "fall" of the Moon. For this, it was necessary to prove that gravity on Earth would remain unchanged if its entire mass were concentrated into its center, which required being able to integrate the individual forces of each tiny element of terrestrial volume upon a body situated on its surface. For Leibniz, it was more a case of the theoretical problem of defining infinitely small or infinitely large numbers. Ironically, the modern version of differentiation, which involves the concept of the limit, was developed in mathematics starting from the Newtonian conception, while Leibniz's notation triumphed among physicists (as it was more descriptive and practical).[8]

What, then, does differential and integral calculus signify for the physicist, and why has its success been so great that we end up identifying this method, and the differential equations we can formulate using it, with physics itself? The answer is perhaps that differential calculus is the practical translation of Descartes' method.

Descartes based his method of scientific analysis on the idea that, in order to solve a complex problem, one needs to separate it into simpler parts, to describe these simpler parts, and then to reintegrate them to obtain an explanation of the whole. This is precisely what differential calculus brings about: instead of attempting to understand all at once how a physical quantity varies from one point to another in space, or from one moment to another, one describes a variation between a point and a point infinitely close to it (or between an instant and one infinitely closely following it). When considered this way, each variation becomes simple. One can then obtain the desired total variation over finite distances and times (that is, no longer infinitesimal) by computing the sum of all these tiny variations (this is integral calculus).

More generally, equations in physics describe simple relations between physical quantities, defined locally by the tendency of their differentiation. Most often, these take the form of differential equations.

The Cartesian method has sometimes been charged with reductionism. This is, in my opinion, a false charge which stems precisely from a reduction of Cartesian thought. Sure enough, a restricted application of this method to a particular problem can be reductive. But it is remarkable that when a blockage of theory arises (as periodically happens in physics), Cartesian reasoning so often leads to a solution to the problem, simply because it is a general method of *analyzing* the laws of nature. For example, the idea that the whole must be the simple and direct sum of its parts (an example of a "reductionist" idea) is not explicitly part of the Cartesian perspective. One must analyze the system in question to find parts that are easier to describe, but the identification of these parts may well be more complicated than a simple "cutting apart," and the reconstruction of the whole more elaborate than a simple "gluing together."

We will see an example of the power of the Cartesian mode of analysis with the theory of scale relativity, where we will relinquish the hypothesis of differentiability. The motivation for this relinquishment is the observation that, in actual experiments, the separation into finer and finer sections of space and time does not necessarily yield parts simpler than the initial system (for example, an electron is simple at a large scale, but at a small scale its internal structure becomes enormously complex, involving the whole spectrum of different elementary particles). At first glance, this point of view would seem to be in disagreement with Descartes: the parts are not more simple than the whole. Nevertheless, a solution can be found to this problem by applying Cartesian analysis at a deeper level. Spacetime itself possesses an internal sub-structure (in the "space" of *scales*) to which differential calculus can be applied, even in the case where it is not differentiable.

The Absolute Space of Newton

If Newton's work founded modern physics in many respects, in particular with the theory of gravitation and its predictive power, it nevertheless blocked the evolution of ideas on another essential point, that of relativity of motion. Newton upheld the existence of an absolute space. This idea, in contradiction with the conceptions of Galileo, Descartes, and Huygens, held sway for over two centuries. If Newton ended up taking such a position, while having explicitly used relativist logic in his construction of the theory of universal gravitation, it was due to an extremely difficult problem, the solution to which is perhaps not complete even today. The problem concerns the apparently absolute character of the forces of inertia which appear in the course of accelerated motion.

To understand what led to a crisis of such duration in physics, one can return to the ship's cabin imagined by Galileo. Galileo had argued that no purely local experiment (inside the cabin and without any reference to the outside) could allow one to determine whether the boat was in motion or at rest, as long as this motion occurred at a uniform and constant speed with respect to the shore. But what about the case of the boat rotating around its own axis? In this case, the appearance of a centrifugal force in the cabin's interior, which the experimenter could detect without looking outside, would seem to imply that rotational movement, contrary to translational movement, can be defined in an absolute manner. Newton used the example of a bucket full of water. When at rest, the surface of water in the bucket is flat; however, as soon as one turns it, the surface becomes curved. One can then, apparently, determine the rest or motion of the bucket "in itself," in a purely "local" fashion, without reference to another object. There would then exist an absolute space: bodies that would be at rest with respect to this space would not be influenced by any force of inertia.

Celebrated contemporaries of Newton were opposed to this vision of the world. For Huygens, well before Poincaré, Mach, or Einstein, there must have been relativity of all motion.[9] For Leibniz, to define a space independently of the objects which it contained could not make any sense. But these few objectors could not prevent the idea of relative space from practically disappearing for two centuries, and the search for ether, also introduced as the medium needed for light waves to propagate, from becoming one of the dominant problems. It would have to wait until the end of the nineteenth century and the beginning of the twentieth, when Poincaré re-examined in detail the relative character of all motion, when Mach proposed a solution to the problem of inertial forces and, finally, when Einstein constructed the theory of general relativity. However, one must not put the cart before the horse: general relativity could not be constructed before special relativity, that is to say before the discovery that space and time are not separate entities, but are in fact subspaces (of three dimensions and one dimension, respectively) or a four-dimensional spacetime.

Chapter 4
Special Relativity

With special relativity, discovered independently by Poincaré and Einstein around the same time, the concept of relativity would strengthen and take on a more profound meaning. Its discovery was necessary: several serious problems affecting the very foundations of physics, even those which had appeared most solid, surfaced at the end of the nineteenth century. These difficulties concerned in particular the relationship between light and motion.

One of these problems was experimental in nature. The work of Fresnel and Young at the beginning of the century, and then the success of Maxwell's theory of electromagnetism firmly established, by the 1880s, the wave-like nature of light. If light was an electromagnetic wave, it seemed inevitable that it needed a medium through which to propagate, which was called "ether." The American physicist Albert Michelson intended to verify the existence of ether by measuring the Earth's movement. However, this measurement proved to be impossible. When Michelson and Morley used interferometry in order to determine the combined speed of light and the Earth moving in its orbit around the Sun, they obtained a strange result. When light was emitted in the opposite direction of the Earth's motion, they expected to find the difference of their speed; in the case where Earth and light travelled in the same direction, they should have found their sum. Yet in both cases they obtained the same result: the speed of light, c, unchanged!

The most solid laws of classical mechanics were thus revealed to be faulty by the experiment. It should seem obvious that, if we walk at the speed of 5 km/h toward the bow of a boat moving at 15 km/h with respect to the Earth, our speed would be 20 km/h with respect to the riverbank. Michelson's experiment yielded a different result: one could add to the speed of light any speed whatsoever, and that speed would stay invariant. It seemed that the laws of addition themselves were put into doubt: 2 + 2 no longer equaled 4! Here was a property which seemed reserved for an infinite speed: only infinity plus another number always equals infinity.

To this enormous experimental problem would quickly be added a theoretical one just as colossal. Oersted, Ampère, and Faraday had understood, since the beginning of the nineteenth century, that electricity and magnetism were seemingly the manifestations of a single field. Oersted called attention to the magnetic field associated with the electric current which runs through a wire. Then Ampère established relations between currents and magnets, which led him to propose a theory of magnetization very close to present-day theory. Faraday went further yet in discovering induction. Currents and magnets have an effect on each other: not only does a current produce a magnetic field (one can create an "artificial" magnet with electric current), but the converse is true: magnets can induce a current. One can thus observe the "direct conversion of magnetism into electricity."[10] Faraday started to develop a theory of an electromagnetic field by using lines of force: such an approach allowed a resolution to the problem of action at a distance, which remained unresolved in the Newtonian theory of gravitation.

But it was Maxwell who succeeded, in the mid-nineteenth century, in constructing a complete theory of the electromagnetic field. It is one of the most beautiful unifications in physics. Not only are electricity and magnetism found to both be manifestations of a single field, but this field can itself propagate in the form of a wave, and the equations of this wave are exactly similar to those which describe the behavior of light waves. Light, then, is identified as one form of electromagnetic wave among others. Inversely, electromagnetic waves with a greater or lesser wavelength than visible light would be able to be manipulated in the same way, meaning they could be emitted, reflected, diffracted, focalized, as demonstrated by the discovery of radio waves by Hertz.

These discoveries would change the world: they led to the invention of the electric motor, the telegraph, the telephone, radio transmissions, and they support all of our modern technology. They also provide a striking example of the connections, often hidden but deep, which tie together basic science and technological innovation. The inventions of a Bell or an Edison would not have been possible without the framework constructed by a Faraday and a Maxwell. The latter and their predecessors themselves could not have constructed electromagnetic theory without the inspiration of the theory of gravitation. The concept of a gravitational field served as model for an electromagnetic field. Previously, Coulomb had directly based his law of electric attraction proportional to $1/r^2$ on Newtonian force. Newton himself constructed his theory on the principles of inertia and additivity of motions put forth by Galileo and Descartes: as he put it, he had "stood on the shoulders of giants." Truly innovative technological progress is not possible without great paradigm shifts within basic science. While it may seem, at first glance, that there is no link between, for example, the television and Newton's theory of universal gravitation, one does exist, and it is indeed a significant one. Fundamental progress, in outlining new ways of thinking, serves as the bedrock without which the edifice of knowledge could not be raised.

Let us return to electromagnetism and its link to the theory of relativity. We may begin by noting that the effects of induction are directly tied to the relativity of motion and show it in a new light, which would be fully explained by Einstein. Indeed, these effects depend solely on relative motion. But if we can understand, or at least admit easily enough, that in moving a body with respect to itself one creates a magnetic field that was not there before this action (after all, one acts on the body by making it go from rest to motion), how is one to understand, on the other hand, that such a field can also appear without one touching the body in any way, simply by deciding to move with respect to it. It is only our state of *relative* motion with respect to the body that determines the "existence" of the field. Let us anticipate the Einsteinian explanation to recall at present the wonderful relativist explanation of this "miracle" of induction. If the magnetic field seems to appear or disappear according to the relative movement between the system and observer, it is in the same way that one can make the face of a cube "disappear" by turning it by 90°. Our experience of rotations of space allows us to know that the face has not truly disappeared, that it is still there, but that, because of the effect of projection, one of its components on the "plane of the sky," to which our eyes only have access, has vanished. It is the same with the magnetic field: this field is only a subset of the electromagnetic field which stays globally unaffected by motion. A motion is nothing else but a rotation in spacetime,[11] a rotation which allows the nullification of some of its components on certain axes and the appearance of other components which had been null in certain frames of reference.

We have come, at last, to the primary reason for which Maxwell's theory played a crucial role in the construction of Einstein's theory of relativity. One of the questions which rapidly becomes evident, once the equations of Maxwell are written in their definitive form, is that of their invariance under the laws of transformation of coordinate systems. It was at the time well established that the laws of mechanics, including gravitation, were invariant under a Galilean transformation (which is the change of coordinates at a constant uniform speed). It was easy to verify that Maxwell's equations were not invariant under this transformation. The task of discovering the new transformation laws which would leave them invariant proved to be more difficult. These laws, in their exact form, were not discovered until 1905, by Henri Poincaré, who named them "Lorentz transformations" in honor of the great Dutch physicist (who had obtained approximating versions). The history of this discovery is worth being told, if only because it is surprising to observe how it is universally attributed to Lorentz, when it is clear, according to his writings, that he never obtained the correct transformation, although he did play an essential role in its discovery.

Poincaré and Relativity

Indeed, one often encounters, at some point in a book or article, a statement to this effect: "Why didn't Poincaré discover relativity? Surely it is because he was too much of a mathematician and not enough of a physicist." Such a judgement is profoundly unjust to Poincaré, for two reasons.

The first is that Poincaré actually did discover relativity, a bit before Einstein. The Lorentz transformation was never established by Lorentz; Poincaré was the first to do so. It is interesting, incidentally, to observe that Poincaré, while having been "banished" from the official history of special relativity, is nevertheless given a permanent place in the everyday practice and vocabulary of today's physicists. It is thus that the complete group of special relativity, that which contains the (so-called) Lorentz transformations (rotations in spacetime) as well as translations, is called the "Poincaré group."

The second is that one needs only to reread him to realize that Poincaré was as extraordinary a physicist as a mathematician (in the tradition of Newton, Euler, and Gauss), intuitive and rigorous at the same time, and that, quite simply, he founded modern physics. While Einstein's and Lorentz's articles on (what would become) special relativity seem (from a mathematical point of view) to still belong to the nineteenth century, those of Poincaré are the first to use the methods of twentieth-century physics: group theory and reasoning tied to the properties of symmetry.

Neither is it a question of arguing that the discovery of special relativity belongs to Poincaré, and not Einstein. There is no doubt that Albert Einstein established the laws of special relativity independently of Poincaré. It seems that Einstein formed his thoughts on relativity in large part by reading the work of Poincaré in 1902, then developed them by himself afterward. He immediately understood its full implications, those which were physically most profound. Thus, Einstein writes in his 1905 article:

The speed of light in our theory plays the role, physically, of an infinitely high speed.

Finally, special relativity was for him just a step in the gigantic construction that he would undertake toward general relativity (as we will see in the following chapter, Einstein posed the principle of equivalence as early as 1907), while Poincaré seems to have underestimated the problem of constructing a relativistic theory of gravitation.

Both discovered special relativity, practically at the same time, and *both* deeply understood its significance to physics, even if their statements vary in form. Here are some examples bearing on the bases of relativistic analysis, which show that Poincaré had effectively taken the step toward the new physics of spacetime (relativity of inertial motion, non-existence of ether, absence of absolute time, relativity of simultaneity). As early as 1899, Poincaré affirmed:

I consider it very likely that optical phenomena depend solely on the relative movements of the material bodies present.[12]

In *Science and Hypothesis*, published in 1902, he writes:

There is no absolute space, and we only conceive of relative motion. . . .
There is no absolute time. . . .
We have not . . . direct intuition of the simultaneity of two events occurring in two different places. . . .
That does not prevent absolute space—that is to say, the point to which we must refer the earth to know if it really does turn round—from having no objective existence.
Whether the ether exists or not matters little . . . [This] hypothesis plays but a secondary rôle. [It] may be sacrificed.[13]

On the 24th of September, 1904, in a speech given to the Congress of Arts and Science in Saint Louis for the Universal

Exposition, Poincaré took stock of the state of mathematical physics at the time and reflected upon its future. The text is a marvel. In it, Poincaré summarizes in a dazzling fashion the nature of the laws of physics, identifies the essential problems that it then faced, makes a diagnosis of the crisis, and recalls the great fundamental principles, identifying those which should survive the transformation in process and those which should disappear or evolve. Finally, he proposes solutions that fulfill an incredible prophecy of what would effectively be twentieth-century physics (even though he had denied doing so at the outset). Below are some extracts:

> There are symptoms of a serious crisis, which would seem to indicate that we may expect presently a transformation. . . .
> This crisis will be salutary . . . [It] is not the first, and in order to understand it, it is well to recall those which have gone before. . . .
> Mathematical physics, as we are well aware, is an offspring of celestial mechanics. . . . Many of us . . . know that the ultimate element of things will not be attained, except by disentangling with patience the complex skein furnished us by our senses. . . . They believe that when we once arrive at these ultimate elements, we shall meet again the majestic simplicity of celestial mechanics.
> [For the ancients, the law] was an internal harmony, statical as it were, and unchangeable. . . . To us a law is no longer that at all; it is a constant relation between the phenomena of today and that of tomorrow; in a word, it is a differential equation.[14]

After powerfully summing up the passage from the ancient to the modern version of science (in which one no longer seeks a pre-existing harmony but certain relations, no longer seeks to establish at once what the real structures are, but to construct equations which can be solved, thus extending the domain of the possible), Poincaré pursues an analysis of what is precisely one of the principal themes of the present work: the scale invariance of many of the empirical laws found in physics, which take the form of power laws.

> Like the stars themselves, [atoms] attract each other or repel, and this attraction or repulsion . . . depends only on the distance. The law according to which this force varies with the distance is perhaps not the law of Newton, but it is analogous thereto: instead of the exponent -2 we probably have another exponent, and from this diversity in the exponents proceeds all the diversity of the physical phenomena. . . . We have the exponent -6 or -5 . . . but it is always an exponent. . . . Such is the primitive conception in its utmost purity.[15]

But this purity is only an arbitrary simplification. The evolution of physics which Poincaré wished to reinforce consists precisely in going beyond this empirical approach, and privileging a deeper understanding based on first principles, even if that would imply giving up the hope of "knowing all":

> Nevertheless there came a day when the conception of central forces appeared no longer to suffice . . . What was done? Abandoned was the thought of exploring the details of the universe . . . and one was content to take as guides certain general principles.[16]

Next, Poincaré lists the great principles of physics: the principle of the conservation of energy, the principle of the conservation of mass (Lavoisier), Carnot's principle of the dissipation of energy, Newton's principle of the equality of action and reaction, the principle of least action, and finally,

> The principle of relativity, according to which the laws of physical phenomena must be the same for a stationary observer as for one carried along in a uniform motion of translation, so that we have no means, and can have none, of

determining whether or not we are being carried along in such a motion.[17]

And he concludes:

> The application of these five or six general principles to the various physical phenomena suffices to teach us what we may reasonably hope to know about them.[18]

With this text, Poincaré founds modern physics. Furthermore, the principle of special relativity is stated there for the first time in its constructive form of invariance of laws under changes of frames of reference, which goes beyond the "passive" Galilean form ("movement is like nothing"), for it allows not only the verification of the relativity of already existing laws (for example the laws of motion and the Galilean and Newtonian dynamics), but also the establishment of new laws. Given the statements made by Poincaré in the rest of this non-specialist text, it is very likely that he had already obtained, in its essential form, the concept of special relativity. He writes in particular:

> From all these results, if they were to be confirmed, would issue a wholly new mechanics which would be characterized above all by this fact, that there could be no velocity greater than that of light.[19]

This work would in fact be published six months later, some time before Einstein's. It is especially remarkable that he speaks here of mechanics and not only electrodynamics: it is thus not for him a case only of finding the invariant group of Maxwell's equations but, as in the work of Einstein, of finding a unique transformation law which could be applied to mechanics in general (not only to the electron) rather than electromagnetism alone.

One must nevertheless temper this success to some degree. Poincaré always considered that the principle of relativity by itself was insufficient in establishing the Lorentz transformation, and that one needed a "complementary hypothesis" (that he chose to be the Lorentz contraction). It is on this point that Abraham Pais makes his case for not completely attributing special relativity to Poincaré in his book *Subtle is the Lord: The Science and the Life of Albert Einstein.* We will see in more detail what this means by analyzing the proofs of Einstein and those more recent. In fact, Einstein had from the outset a larger vision than Poincaré of the principle of relativity, which allowed him to pose certain essential axioms starting from this principle, and then to extend it to general relativity, which Poincaré did not do. He also added a supplementary hypothesis to the principle of Galileo (that of the invariance of the speed of light) and this choice, while equivalent from the mathematical point of view (both use the same number of axioms), would turn out to be much more striking and convincing than that of Poincaré.

In the rest of his paper, Poincaré shines a light on the experimental problems which cast the great principles into doubt. The convergence of his diagnostic with that of Einstein's one year later is remarkable. The four essential problems that he raises can be put into direct correspondence with the four celebrated articles of Einstein in 1905, on Brownian motion, special relativity, the nature of mass, and the photoelectric effect.

Poincaré clarifies first of all the nature of Brownian motion and takes note of its importance with respect to the central question of physics concerning macroscopic irreversibility:

> The biologist, armed with his microscope, has for a long time noticed in his preparations certain irregular motions of small particles in suspension; this is known as Brown's motion. . . . These motions do not cease, or, rather, they come into existence incessantly, without borrowing from any external source of energy. . . . We see before our eyes motion transformed into heat by friction and conversely heat changing into motion, and all without any sort of loss, since the motion continues forever. . . . We need no longer the infinitely keen eye of Maxwell's demon in order to see the

world move backward; our microscope suffices. The larger bodies are bombarded from all sides by the moving atoms, but they do not stir, because these shocks are so numerous that the law of probabilities requires them to compensate each other; but the smaller particles are hit too rarely to have this compensation take place with any degree of certainty and are thus incessantly tossed about.[20]

Thus, for Poincaré, the nature of Brownian motion as due to atomic and molecular shocks was not under any doubt, nor its importance as a crucial experiment showing the limits of thermodynamics. To show this limit was exactly the goal that Einstein gave himself in his article of 1905 (where he developed the quantitative theory). He expressed it in 1915 using words which recall those of Poincaré:

The great significance as a matter of principle is, however, . . . that one sees directly under the microscope part of the heat energy in the form of mechanical energy.[21]

It is equally remarkable that, for Poincaré, the question of reversibility is already a question of scale. One sees, through Brownian motion, reversible laws of microscopic physics leading to irreversible macroscopic laws. Poincaré then addresses the problem of relativity:

All the attempts to measure the velocity of the earth relative to the ether have led to negative results. . . . Michelson carried precision to its utmost limits; nothing came of it.[22]

Poincaré then explains the method of synchronizing clocks (which is also one of the essential reasonings of Einstein):

Watches [A and B] regulated in this way, therefore, will not mark the true time; they will mark what might be called the local time, so that one will gain on the other. It matters little, since we have no means of perceiving it. All the phenomena which take place at *A*, for example, will be behind time, but all just the same amount, and the observer will not notice it since his watch is also behind time; thus, in accordance with the principle of relativity he will have no means of ascertaining whether he is at rest or in absolute motion.[23]

We shall have to construct an entirely new mechanics, which we can only just get a glimpse of, where, the inertia increasing with the velocity, the velocity of light would be a limit beyond which it would be impossible to go.[24]

He continues with the problems encountered by the principles of conservation of mass and energy, in particular the result of the radioactivity experiments of Becquerel and of Pierre and Marie Curie:

The scene changed when Curie thought of placing the radium in a calorimeter. It was then seen that the quantity of heat continuously generated was very considerable. . . .[25]
[The principle of conservation of mass] would cease to be so for bodies having velocities comparable with that of light. Now, such velocities are at present believed to have been realized; the cathode rays and those of radium would seem to be formed of very minute particles or electrons that move with velocities that are no doubt less than that of light, but which appear to be about one tenth or one third of it. . . .
We can no longer distinguish between the mechanical mass and the electrodynamic mass. . . .] The mass can no longer be constant; it increases with the velocity.[26]

As early as 1900, Poincaré moreover understood that one could attribute to a radiation energy *E* an inertia $m = E/c^2$:

If [a] device has a mass of 1kg, and if it emits three million joules in one direction with the velocity of light, the speed of the recoil is 1 cm/sec.[27]

Of course, there is much more in Einstein's work, when he identifies the mass of a solid body with a resting energy.

One last cause of astonishment at the reading of this extraordinary article by Poincaré concerns the final problem he raises:

> [The] dynamics of the electrons can be approached from many sides; but among the roads that lead there, there is one which has been somewhat neglected, and yet it is one of those that promise us the most surprises. It is the motion of electrons that produces the lines of the spectrum. . . . Why are the lines of the spectrum distributed according to a regular law? . . . I believe that here is one of the most important of nature's secrets. . . . Electrons do not behave like the matter with which we are familiar.[28]

Thus it was clear for Poincaré that the spectral lines had an electronic origin (when the structure of the atom was far from having been characterized as nucleus and electron cloud), and that the problem of their quantization (only certain discrete values appeared) would reveal itself as one of the most crucial problems in physics, since it eventually led to quantum mechanics. Here again Einstein made the same observation when he raised the problem of quanta with respect to the photoelectric effect.

Poincaré's conclusions are also prophetic:

> We should have to rebuild from the beginning. . . . Perhaps it is the kinetic theory of gases that will forge ahead and serve as a model for the others. . . . A physical law would then assume an entirely new aspect; it would no longer be merely a differential equation, it would assume the character of a statistical law.[29]

A wonderful vision into the future of the probabilistic character of what would become quantum theory! One might argue that the quantum laws are still written in terms of differential equations, but the rest of this work makes full sense of this intuition.

Of course it might have been that Poincaré only caught a glimpse of the new physics in emergence. It turns out that he did more, and that he was indeed one of the essential authors of its effective construction: he is the first to have established the correct Lorentz transformation, both as an invariant group in electromagnetism as well as in mechanics.

Hendrik Lorentz could have established the transformation which bears his name in an article from 1904. Let us consider this article. Its goal is to respond to an objection of Poincaré to his first works. Lorentz had proposed an approximative transformation law, valid only for lower speeds. Poincaré had noted, in 1900, that only an exact transformation law would be convincing. Lorentz thought, in 1904, that he had finally obtained this transformation. But that which figures in this celebrated article is not what we now call the Lorentz transformation![30] Abraham Pais, who has greatly contributed to restoring Poincaré to his rightful place in the history of relativity (it is thus that Steven Weinberg, discoverer of the electroweak unification, after reading Pais' book, has affirmed that special relativity ought to be attributed to Einstein and Poincaré on a level of equality), nonetheless did not identify this error, and observed only that Lorentz had been mistaken on the transformation law of velocities and could not establish the covariance (now named after Lorentz). He only attributed to Poincaré the correction of the error regarding velocities.

Then who was it who discovered the Lorentz transformation? Henri Poincaré! And why is it called the Lorentz transformation? Because Poincaré himself named it thus in his honor! In a brief paper given to the French Academy of Sciences on June 5, 1905 (thus one month before Einstein's article), he writes:

The results that I have obtained are in agreement on all the important points with those of Lorentz; I have only been led to modify them and to complete them in some points of detail. The essential point, established by Lorentz, is that the electromagnetic field equations are not changed by a certain transformation (that I will call using the name of Lorentz) and which have the following form.[31] The set of all these transformations, together with the set of all spatial rotations, must form a group.[32]

After converting the units and combining the equations of Poincaré, one can see in the transformation given by Poincaré the true "Lorentz transformation." Thus, it was Poincaré himself who, in a gentlemanly manner, attributed the discovery to Lorentz. Lorentz's error seemed to him to be only a point of detail! And yet, this error is not a simple typo and certainly not a point of detail. The reasons for which Lorentz was mistaken, while Poincaré arrived at the correct transformation, lie in the different physical meaning that each gave to this transformation. For Lorentz, it had nothing to do with a problem of relativity, but instead concerned the mechanical contraction of objects, from the fact of their motion with respect to ether. It is doubtless not an accident if his initial error had to do with the transformation of positions and had repercussions on that of time. For Poincaré, on the contrary, the problem posed was to generalize the Galilean transformation. The new transformation must necessarily be aligned with Galileo's for lower velocities, which enabled him immediately to find the correct formula.

Einstein and Special Relativity

A few months after Poincaré, and independently of him, Einstein proposed an extremely deep and thorough construction of the new theory.[33] It was the result of reflections which began almost in his childhood. Einstein had very early and intuitively practiced that mode of investigation (a "thought experiment" that one can call "relativist" *a posteriori*), which consists of placing oneself mentally within the system surrounding the object which one wishes to understand. As early as the age of sixteen, Einstein investigated what an observer propelled by an electromagnetic wave would see. He writes in his autobiographical notes:

> Gradually I despaired of the possibility of discovering the true laws by means of constructive efforts based on known facts. The longer and the more desperately I tried, the more I came to the conviction that only the discovery of a universal formal principle could lead us to assured results. . . . After ten years of reflection such a principle resulted from a paradox upon which I had already hit at the age of sixteen: If I pursue a beam of light with the velocity c (velocity of light in a vacuum), I should observe such a beam of light as an electromagnetic field at rest though spatially oscillating. There seems to be no such thing, however, neither on the basis of experience nor according to Maxwell's equations. . . . One sees that in this paradox the germ of the special relativity theory is already contained.[34]

The "relativist" state of mind is present from the outset for the young Einstein, not only in the methods used, but in the manner of posing the problem. It is a matter of reconciling statements or laws that are apparently contradictory: invariance of the speed of light and the principle of relativity, mechanics and electrodynamics. Thus, for him, the relativist mode of thinking is already at work in constructing a theory of relativity. It is the "philosophic" principle of relativity, requiring the unity of laws (see the second part of the present work), which prevents the physicist from staying in a mode of description where the laws of mechanics and those of electrodynamics are different, if not opposed. And the solution to the dilemma is none other than the relativity of space and time, in a sort of feedback loop of the principle of relativity with itself.

Einstein attempted to clarify the manner in which he finally resolved the problem, aided by a conversation with his friend Michele Besso:

We discussed every aspect of this problem. Then suddenly I understood where the key to this problem lay. . . . An analysis of the concept of time was my solution. . . . Within five weeks the special theory of relativity was completed.[35]

In his 1905 article, Einstein began by stating the principle of relativity by declaring his goal, which was to reunify electrodynamics and mechanics:

For all coordinate systems for which the mechanical equations hold, the equivalent electrodynamical and optical equations hold also.[36]

Then he added another postulate to this principle:

Light is propagated in vacant space, with a velocity c which is independent of the nature of motion of the emitting body.[37]

It is the combination of these two postulates which constitutes Einstein's principle of relativity: the first is that of supposedly universally valid Galilean relativity[38] (for mechanics, but also for the new electrodynamics which was, of course, unknown to Galileo); the second is that of the invariance of the speed of light. Einstein was aware of the shocking character of this proposition, for he immediately added that the second postulate "is at first sight quite irreconcilable with the former one."[39]

The physicist reader of this period could not, in fact, be prevented from thinking that the principle of Galilean relativity should be understood as the Galilean transformation, in which the speed of light is certainly not invariant, and which does not at all concern electrodynamics. The solution to this puzzle is that the Galilean transformation can be generalized, and that the more general transformation, that of Lorentz, satisfies the second postulate.[40]

In the end, it is in the manner of presenting relativity where Einstein differs from Poincaré: both provide an additional axiom, but

Einstein chooses the essential physical postulate that sums up the new theory then establishes it in principle (the invariance of *c*), whereas Poincaré, for whom one axiom merited another, chose an abstract and technical hypothesis (the form of the "Lorentz factor").[41] For Poincaré, nature is organized so that one is never able to observe absolute motion; for Einstein, absolute motion *does not exist*, velocities higher than that of light *lose all meaning*, etc.

One must place oneself into the mindset of the period to fully appreciate the upheaval brought by these new concepts. While the idea of a universal frame of reference defining absolute rest had already been rejected by Galileo and Leibniz, the concept of absolute space had been reintroduced by Newton and constituted the mental framework of physicists for two centuries. Also, the Einsteinian revolution dealt as much with space as with time: Einstein had to reaffirm and prove physically and mathematically that none of the known physical laws, whether they be in mechanics or in electromagnetism, required the existence of a point of reference at rest with respect to which the motion of all others could be determined. The notions of rest and motion can only be relative.

The analysis of Poincaré and Einstein shows that the notion of two event's simultaneity can never be absolute: if two events are simultaneous within one frame of reference, they are not necessarily so in another. In special relativity, time loses its special status and finds itself submitted to the same relativity as spatial positions. When the components along a certain axis of two points coincide, we know well that that does not mean the two points are identical: if one of their coordinates coincides, it is only by an effect of projection. We are accustomed, from the fact of the two-dimensional character of our vision of objects which themselves are three-dimensional, to seeing at all times different objects along the same line of sight. The knowledge which we always have of the fact that space is in actuality in three dimensions and that this coincidence is an effect of projection allows us to correct this illusion continuously. In the same way, temporal coincidences are often nothing other than such effects of projection of events which are not at all identical!

Many consequences of special relativity seemed highly paradoxical at the time of its construction, but can in fact be understood in an intuitive manner if one grasps the deep reasoning. It is thus with the contraction of lengths and the dilation of time. The application of the Lorentz transformation formulae shows in effect that an object moving before us seems to shorten with respect to its length at rest, and that a temporal phenomenon lasts longer within one frame of reference where it is in motion than in its resting reference point. This is not a simple theoretical curiosity: the slowing of clocks is seen all the time in particle accelerators.

The introduction of the concept of spacetime and the four-dimensional mathematical methods which formalize it, by Poincaré in 1905, and then by Minkowski in 1908, makes full sense of the Einsteinian revolution. In classical mechanics, space (in three dimensions) and time (one-dimensional) are two absolute and independent concepts. In special relativity, a four-dimensional entity, spacetime, encapsulates them and this absolute character, so that neither space nor time are any longer absolutes taken individually. It is no longer the element of length dl nor the element of time dt which are invariant within a change of coordinates, but a combination of the two, the four-dimensional Minkowskian "distance" ds, constructed following a generalization of the Pythagorean relation.[42]

The key to understanding special relativity is then quite simple: the Lorentz transformation is a rotation in spacetime. Similar to how a rotation in space leads to a change in the apparent lengths of bodies (by the effect of projection), lengths and times are no longer independent in special relativity, and themselves correspond to a "projection" of the generalized invariant, causing the effects of length contraction. The dilation of intervals of time is of the same nature, but is more difficult to grasp intuitively. In the three-dimensional Euclidean space to which we are accustomed, we see that a ruler always appears smaller after a rotation than if it is seen frontally: there is an apparent contraction of the length projected. But the spacetime of special relativity, if its spatial component is indeed Euclidian, corresponds to a new geometry, called hyperbolic, in what concerns the relations between space and time.

In such a hyperbolic rotation, there is an apparent dilation, not a contraction.

In the case of photons (or more generally of massless particles), the invariance of the speed of light is expressed by the nullification of this element of four-dimensional distance. One understands better the physical significance of the element *ds* if one understands that it is identified with a particle's interval of proper time, defined as the time which passes in the frame of reference where it is at rest, which is directly tied to the particle. It is this property of light of not seeing its own time pass that Einstein had seen at a very young age in an intuitive form, and which was finally realized in the form of an equation once the theory was constructed.

Classical mechanics has established that there exist deep connections between the properties of space and time and the fundamental quantities of energy, linear momentum, and angular momentum. They are called "conservative," since they are conserved, unchanged, over the course of time. Given this, it is not surprising that an evolution of the concept of spacetime implies fundamental consequences for the nature of these quantities. One of the results of special relativity has become known to the point of becoming part of universal culture: the relation of equivalence between mass and energy discovered by Einstein, $E = mc^2$. To establish this relation, one generalizes in special relativity the form taken by the kinetic energy of motion, and one discovers that even at rest, all bodies possess an internal energy equal to the product of their mass by the square of the speed of light.[43]

But work on special relativity and its foundations did not stop at the beginning of the century. Its interpretation was clear for Einstein starting with his 1905 article: it is thus that he writes there, as we have seen, that the speed of light, while finite, possesses within the new framework all the physical properties of an infinite speed. Nevertheless, these foundations have been improved, in particular by the works of Wigner, then Lévy-Leblond, and I have myself proved that one can obtain the special relativity laws using two fewer axioms than the original derivations.[44] This work has shown that the tie between special relativity and the speed of light was only an accident of history. Einstein had obtained the Lorentz transformation by adding to the principle of Galilean relativity that of the invariance of the speed of light. We know today that this supplemental axiom was not necessary, and that the laws of special relativity impose themselves as the most general laws which satisfy the problem posed: that of the construction of inertial transformation laws satisfying the principle of relativity, laws of which those of Galileo are only a special case.[45]

In such an approach, one forgets that one knows (or thinks that one knows) the laws of changing a frame of reference, and one tries to construct them by application of the principle of relativity itself. This principle necessitates that the laws of nature should be valid whatever the system of coordinates. The transformation laws between coordinate systems are part of the fundamental laws of nature: the principle of relativity should also be able to be applied to itself! As a consequence, if one makes two successive transformations of coordinate systems, the composed transformation should be of the same kind as each of the initial transformations.[46]

Similarly, a change in the sign of a spatial or temporal variable should not make any change to the transformation law. It is again a direct consequence of relativity: what is arbitrary in the choice of coordinate systems should in no way influence the physical character of the result.

Yet one can show that the only general transformation of coordinates which satisfies these conditions takes precisely the form of the Lorentz transformation, in the case of transformations of two variables, one of space and the other of time, called *linear*, which are specific to *special* relativity.[47] The constant c does not appear at all in this proof as the speed of light, but as a purely mathematical constant assuring the generality of the obtained transformation. The law of Galileo is equally a solution to the same problem, but in the very special case where this constant would be infinite. Thus the constant c has nothing to do (directly) with the speed of light, no allusion to, or utilization of light coming about in this type of proof. What, then, is the relation and why should the theory have been historically constructed using this connection?

The answer to this question is obtained by studying the new forms that the concepts of energy and linear momentum take in this new theory, compared to Galilean theory. These concepts imply the impossibility for a particle with mass to attain the speed c. It would take an infinite energy and linear momentum for such a particle, no matter how low in mass, to go at the speed c. It is in this sense that this speed is an asymptotic limit, which can never be attained and possesses all the physical properties formerly assigned to infinity. It is thus found to be fundamentally unsurpassable, in the sense that surpassing infinity has no meaning. It is not a barrier, a sort of border in the ordinary sense of the term. In this case, there is no other side of the border; it simply does not exist.

These new laws of energy and momentum have another remarkable property. If it is true that no object possessing mass can attain c, in contrast an object with absolutely zero mass, of which the energy and momentum would be zero in Galilean theory, finds a coherent existence in the theory of Einstein. But such an object is then required to go at the speed c; no other speed is permitted.

To summarize, special relativity implies that material particles can never attain the speed c, and inversely that particles without mass can only travel at the speed c. The correct definition of the speed limit then becomes clear: it is the speed of any massless particle in a vacuum. Light thus does not appear in relativity except as an archetype of a particle having this property. But it shares this property with other particles, such as neutrinos (neutrinos have now been discovered to have a mass, but so small that their velocity remains indistinguishable from that of light). Its role in the construction of the theory comes from what the physicist had available, a physical object which attains the speed c, and which are accessible to him, thus allowing relativistic effects to be shown by evidence. The essential point here is to understand that, even if no physical object could reach this limit (that is to say, if all particles, including photons, were discovered to have mass), this would in no way change special relativity. The speed c

would remain as a speed limit for all transmission of information of any sort.

Ultimately, special relativity demonstrates the existence of spacetime. The complexity of the relativistic formulae compared to their Galilean version is only a consequence of a "bastardized" representation, in which we continue to use three-dimensional concepts such as the usual velocity v, while the transformations are four-dimensional in nature (try to imagine what the representations of movements in ordinary three-dimensional space would be like if one limited oneself to using only two variables). Thus, the finitude of the speed of light is only an effect of *perspective*! It is simply the equivalent of a vanishing point. Here again, one will easily understand it by starting from equivalent effects encountered in our vision of space. Two tracks of a railroad which continue infinitely are seen by us, by reason of perspective, to converge at a finite distance on the "plane of the sky." This comes from the projection of a three-dimensional infinity on the celestial sphere, our vision only being two-dimensional. The maximal speed is similar: to be convinced, it is enough to observe that once expressed in four-dimensional representation, the form of Galilean relations are recovered, in which the speed of light now corresponds to a "four-dimensional speed," which is infinite.

One last question is that of the value of the speed of light. Why is it 300,000 km/s (roughly)?[48] This question is in reality not the right one. As is often the case in relativity, rather than trying to answer it, one must reverse one's point of view. The numerical value of the speed of light is completely dependent on the choice of units. But this is *a priori* arbitrary, and furthermore, the choice of different units of time and speed is only justified if space and time are not connected. Yet special relativity declares that space and time do not exist independent of one another, but are subspaces of four-dimensional spacetime. It is enough to imagine what would happen if we were to measure in space (in three dimensions) the heights in a different unit from the lengths and widths. All would go well until a rotation occurred: one would then see, from the new reference point, the relation between the units complicating all the results of measurement and their interpretation.

It is thus with spacetime and the speed c, which is nothing but an artifact of such a choice of different units: similar to how the relationship between a height over a width is a pure number, without dimension, the true nature of a speed (the ratio of length over time) is also without dimension. In other words, from the moment when we measure space and time with the same unit, which their true nature requires, the speed c disappears from physical laws, becoming the pure number $c = 1$. Since 1985, by decision of the International Bureau of Weights and Measures, there are effectively no longer two separate units of time and length, but only a unit of time. The unit of length is deduced once c is definitely fixed. This method enabled taking into account the relativistic revolution while assuring continuity with the former units. Perhaps one day we will go to the ultimate implications of special relativity and we will measure lengths in nanoseconds?[49]

Speeds would then by definition be non-dimensional quantities and would only be able to vary between -1 and +1. The question of the value of the speed of light would no longer be posed, but only those of the typical values of speeds at which we usually move (for example, 3 m/s would corresponds to 10^{-8}).

Chapter 5
Einstein's General Relativity

In 1907, only two years after developing the theory of special relativity, Einstein had the idea that he would later describe as "the happiest of his entire life." In this (inner) vision, what would be revealed as the essential physical basis of general relativity appeared to him, even if it would take him almost ten years to elaborate the theory mathematically. Einstein realized that "if a man falls freely, he would not feel his weight."[50]

Even the expression "free fall" is telling: though one is apparently always attached to a gravitational field, attracted to the Earth from the perspective of Newtonian theory, one finds freedom when one is falling. It is this freedom that those who pursue free falling as a hobby seek to find and to feel, even if it is only partial due to air resistance. It is of course astronauts in "weightlessness" who truly experience over a long period this feeling of no longer having any weight, of no longer being subject to the force of the Earth's attraction. Nevertheless, the great idea of Einstein was the understanding that, if we jump up, during the brief moment of our jump, we experience this "weightlessness." In other words, there is no difference in principle between a vessel in orbit around the Earth and a ball which we throw here on Earth: both are in free fall; both are, for the duration of their motion, satellites of the Earth.

The Equivalence Principle

Understanding this universal phenomenon led Einstein to formulate the equivalence principle, according to which *a gravitational field is locally equivalent to a field of acceleration*. In order to obtain this principle, he drew upon a fundamental property of gravitational fields already brought to light by Galileo and included in Newton's equations: the acceleration communicated to a body by a gravitational field is independent of its mass.

After the development of special relativity, the need to generalize the theory seemed inevitable for multiple reasons. In fact, relativist unification was far from complete. If the mechanics of free particles and electrodynamics finally satisfied the same laws, it was not the case for Newton's theory of universal gravitation, otherwise the chief showpiece of classical physics. The equations of Newton are invariant under the classical transformation of Galileo, but not under those of Lorentz. Thus physics remained split in two, in contradiction with the principle of relativity, which necessitates the validity of the same fundamental laws in all situations.

Moreover, Newtonian theory is based on certain presuppositions in contradiction with the principle of relativity: it is so with the concept of Newtonian force, which acts at a distance by propagating instantaneously at an infinite speed. The construction of a relativist theory of gravitation thus seemed to Einstein (and other physicists) a logical necessity.

Another problem was just as serious: the relativist approach explicitly gives itself the problem of changes of reference systems and their influence on the form of physical laws. But the answer provided by special relativity is only partial. It only considers frames of reference in uniform translation, at constant speeds with respect to one another. However, the real world constantly shows us rotations and accelerations, from the fact of the multiple forces which are at work (such as gravity), or inversely, causing new forces (such as the forces of inertia).

What are the laws of transformation in the case of accelerated frames of reference? Why would such frames of reference not be as valid for writing the laws of physics as inertial frames of reference? The answer is that such a question requires a generalization of special relativity.

The originality of Einstein's approach had been, in particular, to bring together two problems, that of constructing a relativist theory of gravitation and that of generalizing relativity to non-inertial systems, into a single endeavor. The equivalence principle made this unity of approach possible: if field of acceleration and gravitational field are locally indistinguishable, the two problems of describing changes in the coordinate systems, including those which are accelerated and those which are subject to a gravitational field, boil down to a single problem. But such an approach is not reducible to "making relativist" Newtonian gravitation. While certain physicists could hope, at the time, that the problem of Newton's theory could be solved by a simple reformulation, by introducing a force which propagated at the speed of light, it is the entire framework of classical physics that Einstein proposed to reconstruct with general relativity. Better yet, it was a new type of theory which he developed for the first time: a theory of a framework (curved spacetime, now a dynamic variable) in connection with its contents, and no longer only a theory of "objects" in a rigid preexisting framework (as was Newton's absolute space).

Why such a radical choice? Doubtless because special relativity itself was unsatisfactory on at least one essential point: the spacetime which characterizes it, even if it includes in its description a space and a time which are no longer absolute taken individually, still remains absolute when taken as a four-dimensional "object." However, inspired in particular by the ideas of Ernst Mach, Einstein had come to think that an absolute spacetime could have no physical meaning, but rather, that its geometry should be in correspondence with its material and energetic contents. Thus a reflection on the problem of inertial forces, which had caused Newton to introduce absolute space, led Einstein to the opposite conclusion.

The Problem of Inertial Forces

The existence of inertial forces acutely poses the problem of the absolute or relative nature of motion and, ultimately, of spacetime. The ideas of Mach in this area had a deep influence on Einstein. For Mach, the relativity of motion did not apply solely to uniform motion in translation; rather, all motion of whatever sort was by essence relative (Poincaré and, long before him, Huygens had arrived at the same conclusions).

This proposition can seem in contradiction with the facts. If it is clear, since Galileo, that it is impossible to characterize the state of inertial motion of a body in an absolute manner (only the speed of a body *with respect to another* has physical meaning), it seems different in the case of accelerated motions. Thus, when one considers a body turning about itself, the existence of its rotational motion seems to be able to be felt in a manner totally intrinsic to the body. No other body of reference is needed: it is enough to verify whether or not a centrifugal force appears which has a tendency to deform the rotating body.

In reconsidering the thought experiment of Galileo's ship, the difference between inertial movement and rotational motion becomes heightened. No experiment conducted in the cabin of a ship traveling in uniform and rectilinear motion with respect to the Earth is capable of determining the existence of the boat's movement: as Galileo understood, "motion is like nothing." Relative motion can only be determined by opening a porthole in the cabin and watching the shore pass by. But now, if the boat accelerates or turns about itself, all the objects present in the cabin will be pushed toward the walls. The experimenter will know that there is movement without having to look outside. Thus, accelerated motion seems definable by a purely local experiment.

It is such an argument which caused Newton to allow that one can define an absolute space, in opposition with Leibniz (then Mach) for whom defining a space independently of the objects it contains could not have meaning.

Mach proposed a solution to the problem completely different from Newton's. Starting from the principle of relativity of all motion, he arrived at the natural conclusion that the turning body, within which there appear inertial forces, must turn not with respect to a certain absolute space, but with respect to other material bodies. Which ones? It cannot be close bodies, of which the fluctuations of distribution would provoke observable fluctuations of inertial systems. This is unacceptable, since it is easy to verify the coherence of these systems over great distances. Thus, if we look, motionless with respect to the Earth, at the night sky, we do not see the stars turning.[51] Nevertheless, if we turn about ourselves, we feel our arms spreading out due to inertial forces and, in raising our eyes toward the sky, we can see it turn. This was the initial observation of Mach: it is within the same frame of reference that the arms are raised and the sky turns, and this will be true for two points of the Earth separated by thousands of kilometers. Mach suggested, then, that the common frame of reference is determined by the entirety of distant matter, of bodies "at infinity," of which the cumulative gravitational influence would be at the origin of inertial forces. In other words, the body would turn with respect to a frame of reference, not absolute, but *universal*. An absolute motion would be defined in itself, independently of all objects. However, Mach argued, all motion is relative, remaining defined *with respect to* an "object," even if this object is the universe in its entirety.

The solution proposed by Einstein, that of the equivalence principle and general relativity, incorporates some of these ideas while ultimately distancing himself from the principle of Mach, even though his premises were identical. The distribution of matter and energy in the whole of the universe determines the geometric structure of spacetime, and then the movements of bodies are brought about within the framework of this geometry tied to matter.

Relativity of Gravity

Let us now return to Einstein's great idea in 1907. If an observer descends in free fall within a gravitational field, they no longer feel their weight, which means that they no longer feel the existence of this field itself. This remark, which can now seem obvious to us—we have all seen, on television or in movies, astronauts in weightlessness floating in their ship, and the objects that they drop going away from them at a constant speed—is nevertheless revolutionary, for it implies that gravity does not exist in itself, that its very existence depends on the choice of a frame of reference.

He thus distanced himself from the former concept of gravity. What, apparently, can be more absolute than a gravitational field in the Newtonian model? Gravity had been recognized by Newton as universal; here indeed was a physical phenomenon of which the existence does not seem to be able to depend on such and such a condition of observation.

However, if we allow an enclosed area to fall freely within a gravitational field, and then put in motion a body at a certain velocity with respect to this area, the body will move in a straight line at a constant speed with respect to the walls of the enclosure; a body initially immobile (again, with respect to the walls) will stay thus during the movement of the enclosure's fall. In other words, all experiments that we can perform there would confirm that we are in an *inertial* frame of reference! Thus, gravity, however universal it is, can be cancelled out solely by a judicious choice of coordinate system: what Einstein understood in 1907 was that even the existence of gravity was *relative* to the choice of coordinate system.

The principle of equivalence can thus be understood as a principle of the relativity of gravity. Gravity no longer exists "in itself," but is manifested from the fact of our belonging to certain coordinate systems. If we feel gravity on Earth, it is no longer, in Einstein's theory, because a "force" attracts us toward the center of the Earth, but because the existence of the terrestrial ground requires us to remain fixed in the "wrong" coordinate system. It is enough to place oneself within the appropriate system (that which is in free fall in the gravitational field) to see this field disappear again.

In actuality, the fall of bodies is only illusory. If we throw a body in front of us, it will effectively approach the ground until the body strikes it. But the final shock comes from the existence of the ground, which is not created by gravity alone. The movement of the body before the shock would be almost unchanged if all the mass of the Earth was concentrated in its center. This thought experiment allows one to elucidate the true nature of a body's motion. Without the ground to stop it, it would pursue its movement toward the center, then move further away to the opposite side, and would finish by returning to its point of departure to start its journey again. In other words, the body would follow an elliptical path of which the center of the Earth would be a focal point. When we throw a body, we are doing nothing less than creating a satellite around the Earth.

Everything occurs as if Einstein had taken the propositions of Newton backward. To the original question, why do bodies fall on Earth while the Moon does not fall?, Newton answered that in fact the Moon is always falling if we compare its movement to what it should be (rectilinear and uniform), if it were free of all force. Einstein answered it the other way around: *neither the Moon nor the body falls; both are satellites of the Earth.* It is of course the definition of what we consider to be "falling" which explains the difference. These answers both allow the construction of a universal theory of gravitation, but that of Einstein was shown to be deeper.

The reverse consequences of the equivalence principle are equally true. The conditions of life in a ship uniformly accelerated at 10 m/s^2 would be completely identical to those we know on Earth: we would weigh the same, an object thrown would follow the same parabola. The various experiments that we could conduct would not allow us to distinguish between this field of acceleration and a field of gravitation.

This is of course only true "locally," that is, in a small region with respect to the extent of the field. In a very large vessel, one would be able to verify that the acceleration of the artificial weight stays parallel in a single direction. By contrast, within a gravitational field created by a mass, acceleration is central: precise measurements between different points far apart would show the non-parallelism of the force of gravity. For a large object, tidal forces would appear, deforming the object by the fact of the *variation* of the gravitational force between its extremities.

Generalizing the Principle of Relativity

The two problems which Einstein posed for himself, to find a relativist theory of gravitation and to generalize special relativity to non-inertial frames of reference, thus find themselves indissolubly connected by the equivalence principle. The sole fact of wanting to write the laws of physics within systems undergoing whatever kind of motion, and in particular uniformly accelerated movements, will imply that, in certain frames of reference, gravitational properties will appear in the absence of mass, while in other frames of reference, even in the presence of a mass, one will have all the properties of an inertial system.

Einstein's principle of general relativity therefore extends the principle of relativity to arbitrary movements. It supposes that the laws of nature hold true, whatever the state of motion of the coordinate system. In other words, all systems of reference should be equivalent for the expression of physical laws, whatever their state of motion.

The principle of general relativity is mathematically translated to another principle, that of *generalized covariance*. This principle necessitates that the *form* of physical equations stays unchanged within whatever transformation of coordinate systems (that is to say, in actuality, that they will keep the same form that they had in an inertial system, as we will see in the second part of this work).

Relativity of Geometry

The link between geometry and gravity can be further strengthened. If gravity is relative to the choice of coordinate systems, it is the same for certain geometric statements, as we shall see.

From the point of view of an external observer, a space station in orbit around the Earth periodically moves in ellipses around it. But for the astronaut inside the station, as long as they do not look out the window, everything that they can see in their environment seems to show that they are truly situated in an inertial frame of reference: an object dropped without initial velocity floats without moving; if they push it, it will move at constant velocity until it hits a wall. Let us suppose that the astronaut throws an object very gently and it crosses the station during the time of a fraction of its orbital period. Within the vessel, the object follows a rectilinear trajectory at constant velocity. But what would its trajectory be for a terrestrial observer? An ellipse, that is, a curved trajectory. Thus, the presence of a gravitational field is reflected by the fact that the straight line within one frame of reference is a curve in another.

Another observation made by Einstein, his choice of non-Euclidean geometries, was even more crucial. That one must abandon Euclidean geometry is in fact a direct consequence of special relativity. Let us consider the measurement of a disc's circumference at various distances from its center. The relation of the circumference over the radius of a circle drawn on this disc will always be 2π if it is at rest with respect to the ground.

But let us now consider a disc that turns. One might seek to find out, if one does not yet make use of a more general theory, what special relativity can tell us about such a movement. If the disc turns about itself at a uniform speed, the speed of translation tied to this rotation grows with the distance from the center. By now measuring the circumference with enough precision (or if the disc turns so fast that the speed of its outer rim approaches that of light), its length is diminished by the relativistic contraction of Lorentz, and this contraction increases with the linear speed, that is, when the distance to the center grows. Thus, the relation between circumference and radius of a circle drawn on a turning disc only equals 2π at the center and *decreases* toward the outside. Yet the value of this relation is one of the bases of Euclidean geometry, which thus finds itself overturned by relativity.

Finally, because neither gravity nor the geometry of space (and of time) are absolute, and because the absence of geometric effects (rectilinear motion) corresponds precisely to the absence of gravity (inertia), Einstein was led to posit that gravity and geometry are ultimately one single, identical concept. The theory of relativity had finally arrived at a statement precisely of this kind, in *identifying the effects of gravity as the manifestations of the curvature of spacetime.*

Curved Spacetime

Once the idea of the equivalence principle and that of a non-absolute spacetime depending on its contents has been accepted, a new geometric tool becomes necessary to put in effect the theory of general relativity. Clearly, neither Euclidean geometry, nor its extension in special relativity, the spacetime of Minkowski, are suited for such a project. Both are rigid, absolute, and do not contain the structures which permit one to tie together matter and geometry.

In Euclidean geometry, the set of points situated at a constant distance from a fixed point is a circle in a plane, a sphere in three-dimensional space, a hypersphere beyond that. The circumference of a circle of radius r is $2\pi r$, its surface πr^2, the surface of a sphere $4\pi r^2$, and its volume $(4/3)\pi r^3$, etc. The sum of the angles of a triangle is 180°, and two parallel lines never cross.

Let us now place ourselves on the surface of a sphere, and imagine that we are two-dimensional beings constrained to living on this sphere. Suppose that not only us, but all physical objects propagate themselves exclusively on this sphere. When we speak of "dimension," of "axes," of going in a "straight" line, we make reference to the only thing we know: moving within our "universe," which we cannot know, by purely local measurements, is the surface of a sphere. Not being able to extract ourselves or to observe or make use of anything outside of it, we will be informed of the nature of the world only by its intrinsic, internal properties. Yet measurements at large scale will show that circles there have a circumference less than $2\pi r$ and a surface less than πr^2, that the sum of the angles of a triangle are greater than 180°, and that parallel lines can cross. This "universe" is spherical and corresponds to the surface of a sphere.

Let us return to our own universe. The surface of a sphere can thus be seen as an object (in two dimensions) immersed in a three-dimensional Euclidean space, but also from the *intrinsic* point of view like a genuine two-dimensional space. One can make measurements of length and establish relations between these measurements by exclusively using two coordinates defined along this sphere, and no change of reference will allow one to eliminate the difference between our observed relations and the relations obtained within a plane.

These observations have led to a profound paradigm shift, to the creation of a new geometric universe. It is Gauss, Bolyai, and Lobachevsky who deserve the credit for having truly understood and accepted, at the beginning of the nineteenth century, the logical possibility of a non-Euclidean geometry. Gauss discovered that the methods that he had developed for intrinsically describing the sphere allowed him to describe another two-dimensional geometry, completely new, coherent, and that, nevertheless, *one could not construct as a surface immersed in our usual Euclidean space*. This space, called *hyperbolic*, is infinite and possesses all the attributes of an authentic geometry, which allows one to compare it to Euclidean space: one can construct "lines" of infinite length, and define an infinite number of parallel lines, which we cannot do on the sphere. But parallel lines can cross, the sum of angles of a triangle is less than 180°, the circumference of a circle greater than $2\pi r$. It is a full-fledged geometry, but one which no longer satisfies the axioms of Euclid.

These new geometries are characterized by a new property, curvature. The sphere is a two-dimensional space with constant positive curvature, given by the inverse square of its radius. One can equally envision surfaces of negative curvature: the surface of a horse's saddle, which curves upward or downward in orthogonal planes, is a good example of this. The hyperbolic space of Gauss is precisely a space where the curvature is negative but, moreover, *constant*.

Once the possibility of a geometry not satisfying all the axioms of Euclid was understood, one still had explicitly to construct it, to make a choice among all the possibilities available to the imagination. Gauss decided to limit himself to *locally Euclidean* spaces, that is, for which the law of Pythagoras remained valid for an infinitesimal area around each point. This choice, conserved by Riemann in his generalization to spaces having an arbitrary number of dimensions, will be seen as essential within the construction of general relativity by Einstein: this is the locally flat character of space (more precisely the locally Minkowskian character of spacetime) which will allow the mathematical expression of the principle of equivalence.

Let us observe at this juncture that the theory of scale relativity, which we will speak of later, consists precisely in abandoning this hypothesis. Spacetime is no longer locally flat, but fractal, which means that it possesses internal structures even in small scales.

Gauss had studied the metric properties of curved spaces, such as for example the relations between the lengths of the sides of a triangle. He discovered that these properties were unique to the spaces and allowed one to distinguish between them, with respect to each other as well as with respect to flat space. The internal properties of these spaces boil down to the knowledge of the *distance* between two arbitrary points.

The requirement of covariance (of being able to write the laws of physics under the same form whatever the coordinate system) is achieved when one knows how all the physico-mathematical quantities defined on a space are transformed, and when one can use this knowledge to write equations of which the form no longer depends on the particular system of coordinates chosen (only numerical values taken by the quantities will depend on the choice).

Geodesics

The theory to be constructed ought to take into account the trajectories of bodies. The laws of motions always boil down, as a final resort, to a process of optimization: traveling the shortest path. In the case of ordinary, Euclidean space, the shortest trajectory between two points is a straight line: thus inertial Galilean movement, uniform and rectilinear, answers well to such an optimization. The principle of general relativity requires that all frames of references be equivalent in the description of motion. It must then be that the trajectory of the free particle in a gravitational field follows the equivalent of a "straight line," but within a curved spacetime. What is the equivalent of the straight line in Riemannian geometry? It is what we call the *geodesic* of space. For example, the geodesic upon a sphere, well known to navigators, is composed not by an arc of constant latitude, but by a "great circle" (see figure 1).[52]

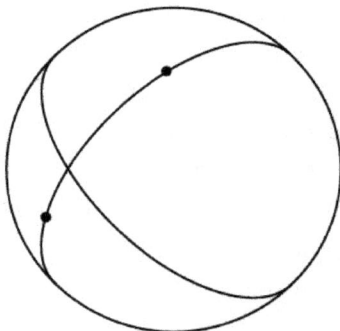

Figure 1 *Geodesic on the surface of a sphere.* The shortest line between two points is an arc of a great circle, that is, an equator passing through these two points.

Inertial motion and free fall within a gravitational field become the same thing: in both cases, they are free movement. This extraordinary result is reflected in the equation of the geodesics, which has exactly the same form, generalized within a curved space, as the equation of inertial motion! It is expressed simply by writing that acceleration is zero, and thus that the object is moving at a uniform constant velocity.

But this acceleration that is canceled out is no longer only the usual acceleration. It contains in its definition the effects induced by geometry on movement, which one describes with the help of a generalization of ordinary differentiation, the covariant derivative. Ultimately, these effects simply demonstrate the general relativity of motion.

Let us consider a physical vector magnitude, like a velocity or a force. It is represented by a small arrow—a vector. Let us now move the origin of this arrow. If, in the course of such a movement, the arrow remains unchanged, parallel to itself and of the same length, then it could have been defined in an absolute manner. Inversely, abandoning the absolute character of spacetime implies that, when being displaced, a vector should turn. What the new derivative defines, and what Einstein's theory of gravitation is, is exactly that: an effect of rotation induced by translations. There is no need to find a "cause" for such an effect. It is the simple expression of a larger generality, in the sense in which it would be its non-existence that would require a cause (owing to the non-absolute character of spacetime).

Thus, Einstein arrived at statements concerning gravity just as powerful as those obtained by Galileo for inertial motion. "Motion is like nothing," it has no need of a cause, it is rest or the variation of motion which requires one. Similarly, spacetime is, *a priori*, curved, for if it was not, this would be equivalent to making the arbitrary hypothesis of its flatness and thus its absolute character. Otherwise said, *gravity is like nothing*, it is as if it does not exist. This property is shown effectively in the fact that an incalculable number of *curved* geometries are solutions to Einstein's equations *without source terms*, that is, without matter or energy (we will see below that matter acts as source for gravity, in the same way that electrical charges are the sources of electromagnetic fields). If curvature and matter are indeed connected by general relativity, it would be misguided to reduce this connection to simple relations of cause and effect, since there can exist curvature without matter. It would in this case be a pure gravitational field, without cause other than the greatest generality of possibilities.

Riemannian geometry would allow Einstein to construct a one-of-its-kind theory in physics: all the usual and formerly indispensable concepts, of force, of potential, even of field, disappear in favor of the sole concept of spacetime. In general relativity, matter changes the curvature of spacetime and free test-particles follow the geodesics of this curved spacetime.

The Sources of Gravity

One final stone of the foundations is still missing: if spacetime is now curved and fluctuating and no longer rigid and immovable, what is the connection between this curvature and matter? What is it that curves spacetime? The nature of field theories is to define sources for the fields, which are the magnitudes that create them. More precisely, one introduces *charges* (such as the electric charge in electricity, or the strong or "color" charge in the theory of strong interaction), then one considers the movement of these charges, which is their *current* (it is thus with electric current). Finally, the equations of a field consist in expressing that the current determines the variations of the field.

For gravity, part of the answer can be found in Newtonian theory: in this theory, the attractive force which deviates trajectories of particles is provoked by *masses*. Should one deduce that it is masses which curve spacetime? The theory of special relativity teaches us that such a response must be insufficient. Einstein demonstrated that mass boiled down to a rest energy (this is what is expressed in the relation $E = mc^2$) and that energy itself is only the temporal component of a more complex mathematical entity in four dimensions mixing energy and momentum (as is the case with time and space). One should speak no longer of separate energy and momentum, but of energy-momentum.

In the theory of Einstein, the "charge" for gravity (its "source") is then energy-momentum in all its forms, that of matter but also that of radiation. Under what form will this "charge" appear in the final theory? We know that, in the equations of electromagnetism, one sees the *electric current* appear as "source" of the field, which represents the movements of electric charges (the product of charges times their speed). The physical action of a field is provoked by the current of acting charges. It will be the same for gravitation, but with a very special result in this case. The equations of gravity will in effect depend on the "current of gravitational charge," that is to say, on the *current of energy-momentum*, while momentum (product of mass times velocity) is already a "current of mass." It is this extraordinary physical quantity (a "current of current") which is identified in Einstein's theory with the geometry of spacetime.[53]

Finally, in the theory of general relativity, mass as well as emitted rays play an active and passive gravitational role: for example, the electromagnetic field contributes, with masses and all other forms of energy, to curve spacetime and, in return, masses, photons, and all other particles and bodies see their trajectories deviated by gravitational fields.

Spacetime Equations

The final step in the construction of the theory consists of determining what geometric expression can be identified with this current of gravitational charge. The equations sought by Einstein should be written in the form GEOMETRY = MATTER. Thus, the two terms of this identity must have the same properties. Yet matter is characterized by a conservation equation, which expresses its continuity (in spacetime) in a generalized manner. Does there exist an equivalent geometric property?

Riemannian spacetimes are, effectively, universally characterized by the fact that "a border has no borders." For example, the border of a square is its perimeter. Upon this, a moving point could travel without ever encountering an obstacle.[54] The same goes for a cube. Its border is constituted by its six faces, which form a finite and closed space in two dimensions, equivalent to the surface of a sphere. In a space with four dimensions, this property becomes more difficult to visualize. The border becomes the three-dimensional "wall" of a hypercube. This wall is made up of eight cubes, each connected to the others by their respective faces. A three-dimensional being like ourselves, who lived within this wall, would never encounter any border, even though such a space would be finite. Here is a general property of continuity and of conservation that is purely geometric, which can then be identified with the equivalent property of continuity for matter: such an identity is what the equations of spacetime in general relativity boil down to.

Finally, what is the nature of Einstein's equations? One can see them as the expression of a constraint on the universes which are physically possible. If all curved geometries are mathematically coherent, they are not so from the point of view of physics. Physical geometries must in fact be connected to the distribution of matter *via* these equations. But this does not prevent there being an enormous diversity of possible solutions. Due to the complexity of Einstein's equations, only a few exact solutions are known. These are extremely particular solutions, which one can only obtain using highly simplifying hypotheses. It is like this for the cosmological solutions describing the universe in its entirety using the hypothesis of homogeneity and isotropy at large scale, or for the solution of Schwarzschild, based on the hypothesis of spherical symmetry around a mass, which describes the behavior of planets around the Sun. But as soon as one arrives at more complex configurations, such as, for example, the problem of the movement of a binary star system, there is no longer any exact known solution (we do, of course, have approximative and/or numerical solutions).

These equations also achieve the goal that Einstein had set, that of a general relativity of motion. Effectively, these are the most general equations from among the simplest that one can write, which would be invariant under changes of coordinate systems that are continuous and twice differentiable (i.e., one can define a speed and an acceleration). Gravitation in Einstein's theory becomes the set of manifestations of spacetime's curvature. This curvature is itself the expression of a greater generality (in relation to the hypothesis according to which spacetime would be flat).

One often tries to express the essence of general relativity by stating that "matter curves spacetime." This mode of description is not quite exact, knowing that curvature can exist without matter and that, inversely, a great quantity of matter can correspond to a spatially flat spacetime. It is because of this that numerous solutions of Einstein's equations for a vacuum exist, of which the most simple are what we call gravitational waves (these are fluctuations of geometry which propagate themselves in a wavelike manner).[55] All the cosmological models, even those which are at the limit of zero density, are spatiotemporally curved, which is translated by the phenomenon of the expansion of the universe, meaning the dilation of their spatial part over the course of time. Regarding this dilation, spatially flat models correspond to a density, called critical, already quite high. Below this density, curvature is negative, above, it is positive. All these examples show that it would be more correct to translate the nature of the theory by the statement that *matter and energy change the curvature of spacetime*. Once again, it is the difference which is important (here of curvature), not curvature in the absolute. One must not then interpret the relation between matter and geometry in the equations of Einstein as "causal" (one would "place" matter which would "then" curve spacetime), but as being an identity (in each point and at every moment, the two members of these equations, one geometric and the other material, are identical).

Thus the observed structures of our world would be the most general of physically possible structures; no other world would be imaginable. To arrive at a description yet more general, one would have to remove the last residual hypotheses of general relativity: differentiability (which is the project of the theory of scale relativity, of which I will speak at greater length later in this work) and, beyond, possibly, of continuity (if this can have physical meaning, which I personally doubt in the framework of a theory of relativity).

Another essential property of general relativity, which it shares with no other theory, is the connection it allows between equations of the "gravitational field" (that is to say of the geometry of spacetime) and equations of particles moving within the field. For example, in the Maxwell-Lorentz theory of electrodynamics, one must add to the electromagnetic field equations those of the Lorentz force, which acts upon particles within the influence of this field. In Einstein's theory, the equations of motion are no longer independent of those of the field, but are deduced from them. This unique property demonstrates the power of the spatiotemporal and geometric approach. It comes from the fact that the trajectories of particles are no longer seen as the result of the action of a force, but tied to the geometric structure of spacetime, defined by its geodesics.

I emphasize, to conclude this first part, the complementarity of the three great principles used by Einstein to construct his theory. The generalized *principle of relativity* states that motion and gravity are always relative to the choice of the coordinate system. The non-existence in itself of gravitation is derived from this. The strong *principle of covariance* requires that the laws of motion within a gravitational field (the equations of geodesics) be able to be written under the form that they have within a vacuum lacking all force, that of the inertial laws. Finally, the *principle of equivalence* specifies definitively that this locally inertial reference frame is identical to the uniformly accelerated coordinate system, in free fall in the gravitational field. These three principles are thus revealed as three ways of illuminating the same whole, which is the connection: relativity—non-existence in itself (of gravity)—free motion.

Part Two
The Principle of Relativity

Chapter 6
What the Principle of Relativity Means

The principle of relativity postulates that *the fundamental laws of nature are valid in any system of reference, whatever its state.*

Having followed the evolution of this principle and its application (we have seen that theories of relativity underwent many successive levels of elaboration over the course of centuries), we will now try to analyze the true meaning of this statement. In this form, it generalizes the meaning Einstein attached to it in 1916: for Galileo, then Poincaré and Einstein, the statement only concerned relativity of *motion*.[56]

The state changes of coordinate systems taken into account in the relativity of today are those of position, orientation, and motion. But the point of view developed in the present work is that *a much broader sense can be given to the principle of relativity, due to an extension of the concept of a reference system*: relativity, in this broader sense, is no longer a fixed and finished principle, but an evolving one. Lévy-Leblond has already insisted on the necessity of making a distinction between the principle of relativity and the theories that one can construct based on it. This distinction is all the more true when considering the most general version of the principle, as we will do here. Our idea is that the principle of relativity applies not only to motion, but to all the magnitudes which characterize the state (which is *always relative*) of the reference system.

To reduce the principle of relativity to existing theories of relativity, theories which remain incomplete, would be to kill the goose that lays the golden eggs! Indeed, it is the "philosophical" principle of relativity in the most profound sense that we will now analyze in more detail.

The Laws of Nature

The principle of relativity is a statement which concerns "the fundamental laws of nature," in the most general sense of the term. From this point of view, this principle can be considered in essence as belonging more to philosophy than to physics. What are the "laws of nature"? One cannot expect a definition here: the goal of this chapter is precisely to show how this concept has transformed over the course of evolution of physics. Laws once thought fundamental have been relegated to a secondary status in favor of more profound laws. As an example, the three laws of Kepler on the motion of planets later become consequences of Newton's law of universal gravitation; today this latter law can be deduced as an approximation of Einstein's general relativity.

Do there even exist laws of nature? Their existence is a foundational hypothesis, underlying physics and, more generally, all scientific knowledge. It is a presupposition necessary to the scientific endeavor. The progress of science, our furthering of knowledge, the successes of experimental results, all support this hypothesis, and demonstrate its effectiveness, but they cannot prove it.

The physicist, in practice, uses only the equations of physics. These equations, one hopes, approach more and more closely these ultimate laws, pinning them down with greater and greater precision. Some physicists have even asked whether there will be an eventual end to physics: will our equations one day express these laws in a perfect manner?[57] Notably, such ideas arise regularly, particularly during periods of crisis (latent and often inexplicit), like the end of the nineteenth century or the current time period.

Another perspective is possible: our equations will only ever be imperfect approximations, valid in a limited framework and always restricted by a border (expanding over the course of time), and past this border, the representation that they give of the world becomes more and more inexact. By viewing the evolution of science this way, the search for knowledge could well have no end. These (hypothetical) laws of nature would only be an asymptotic concept, a horizon, always glimpsed but never attained. It is the dynamic change which counts, the movement of the mind, the path which one follows, and not the end, which could be, in itself, just an illusion.[58]

But this way of thinking should not be misappropriated (as is the case in certain anti-scientific discourses which are currently flourishing): in no way does it mean that science "is wrong," nor that scientific truth does not exist! This border, which limits the domain of a theory's reliable applicability at a certain time, is expanded much further by later theories, and in this process all the successes and knowledge gained through the preceding theory are conserved. A good example of this is the way in which Einstein's general relativity encompasses Newtonian theory in an enlarged framework, and in this framework one still finds all the extremely precise results of Newton's theory. But Newtonian description, so precise as long as the gravitational field stays weak (as it is for the gravity on the surface of the Earth, or for the motion of planets further from the Sun), possesses its own border. Past this border, observations contradict its predictions (as with the motion of Mercury, the closest planet to the Sun, as well as light rays closely passing the solar boundary). It is this particular relation between the two theories which sometimes leads people to consider that Einstein's contribution only brings "relativistic corrections" to Newtonian theory (one speaks, for example, of the *advance* of Mercury's perihelion), while objectively it is the entirety of the phenomenon considered that general relativity predicts (in our example, the position of Mercury at every instant since its discovery).

Thus, in this evolutionary dynamic of science, continually opening new fronts of research, certain laws become established in a definitive manner, and they remain so in later expansions of the paradigm: this is the case, for example, of extremely fundamental laws like the unsurpassability of the speed of light in a vacuum, or the existence of an absolute zero for temperature.

Toward More and More Fundamental Laws

To better grasp the nature of "truth" in science, one must understand that the successive enlargements of one's frame of mind as physics has developed are accompanied by a broadening of the concept of a "law of nature." This is an essential point which, if not clarified, can lead to endless misunderstandings and, eventually, to scientific dialogues carried out as if between deaf persons. Laws which appear as fundamental in a certain time period are shown, in an enlarged framework, to be nothing more than phenomenological laws (that is, restricted to particular phenomena). The example already given of the relation between the laws of Kepler and those of Newton will allow us to better understand. Kepler's three laws are purely descriptive.[59] They concern geometric properties, discovered empirically, of planetary orbits. They are attached to the phenomenon to be described (the global motion of the planets), but do not provide any explanation. With Newton's law of universal gravitation, physics goes a step further. One no longer is concerned with describing the result (the shape and speed of orbits), but the source of motion (the Newtonian force at work between bodies). Kepler's laws can then be shown as consequences of this more fundamental law, and are revealed to be a simple subset of every kind of motion within gravitational fields.

Thus, within a new frame of mind, more general and profound laws emerge, starting from which the preceding laws, often specific to a given reference system, can be derived as a special case. Here is one of the key principles used for resolving the greatest contradictions ever encountered in the pursuit of knowledge. It is by such leaps that physics truly advances. Moreover, when the mental framework expands, apparently unsolvable problems are more often sidestepped than resolved; in other words, the paradigm shift reveals that the questions were in fact poorly formed, since they came from a limited perspective. It is useless to try to answer bad questions. It is better to formulate them well, and indeed, just this work of reformulation is often sufficient to find the answers.

Numerous examples of this have been given in the first part, which very often manifest the principle of relativity and its unifying power. Consider once more the Newtonian unification: what had been considered before the intervention of Galileo and of Newton as the "laws" of terrestrial motion and celestial "laws," laws apparently opposed to each other and contradictory (terrestrial bodies always fall, while celestial bodies perpetually stay in the sky), become manifestations within different coordinate systems of a single, more profound law, that of universal gravitation. Einstein made a similar leap two-and-a-half centuries later. Gravity itself became one of the manifestations of the general relativity of motion. Geometry and matter became unified, and in this larger frame of mind, the law of gravitation, which Newton had set out in an empirical manner, became a derived result.

Relativity as Unifying Principle

What does the principle of relativity tell us about the laws of nature? That they are unique. That they apply in all cases and in every situation. That there is not one law here and another there, one law for certain conditions and another if the circumstances are different. The principle of relativity is, indeed, a great unifying principle. What it says is that the world is one.

But from the moment when one has admitted that there exist laws of nature (and that one understands by that *fundamental* laws), such a statement becomes a matter of logic. In fact, to better understand the principle of relativity, the best way is to imagine what a world disobeying it would be like: a universe where different laws would prevail in different conditions would simply be a world without law.

The statement even of the existence of laws in a fundamental sense necessitates their universality, and thus their applicability to all systems.

One could then push the argument to the point of stating that, ultimately, the principle of relativity does not exist as such, and boils down to a tautology! The statement of the existence of laws, universal by nature, is sufficient in itself. It is the logic of the world's organization that requires it. Said otherwise, the principle of relativity is reduced to the basic postulate upon which science is founded:

There exist laws of nature.

Can we go further? Is this postulate itself justifiable? One of the most effective measures, already used above, of understanding the meaning of a statement is to analyze its opposite. To understand the statement of the existence of laws, one must ask oneself what a world without law would be like. One can already observe that the calculation of probabilities teaches us that statistical laws arise in the absence of law (from pure chance). Wherever chance rules, the laws of chance reign. A typical example of this is thermodynamics, with its laws constructed based on statistical physics. It is precisely upon the random character of the motion of individual particles that new, global laws are constructed, leading to the emergence of thermodynamic magnitudes such as temperature and pressure. One could then propose the paradoxical statement according to which, in a world without law, laws would probably appear due to the mere statement that there are no laws.[60] A world without any laws seems difficult to imagine.

Thus, among the fundamental laws, the principle of relativity itself stands out as the most fundamental law. Here is a working hypothesis, an extremely fecund and powerful heuristic principle, which is in no way reductive or contradictory with the eventual absence of limits in the search for knowledge that we spoke of above: in effect, and I can never insist enough upon this point, the principle of relativity is a dynamic principle, which contains in itself its own evolution.

Principle of the world's oneness, relativity has effectively intervened at essential stages of the history of physics to reunify a disconnected world. It can act as a sure diagnosis of a crisis (do the same laws apply everywhere?) and can also supply the remedy (to take into consideration still more general reference systems). The evolution of scientific ideas is punctuated by great paradigm shifts in which the world has been reconstructed and different universes reconciled thanks to it. Let us recall a few examples, already mentioned in the first part of this work.

The pre-Copernican vision of the world was profoundly anti-relativist, with its center of the universe and its terrestrial laws in contradiction with celestial ones. The new system upheld by Galileo unified these separate worlds.

It would be similar in the case of Poincaré and Einstein's special relativity: before its construction, physics had been cut in two, with the laws of mechanics on one side, regulated by the Galilean transformation, and the laws of electromagnetism and radiation which did not satisfy the transformation. Two classes of objects, light and matter, two types of different fundamental laws? Relativity necessitates unique laws; one of the two laws would be false or incomplete. Only one law would triumph (the Lorentz transformations), and light and matter found themselves united. This unity of the law of change of reference systems necessitates, in addition, the unification of space and time into a single concept, spacetime, then similarly unifies mass and energy.

After the advent of special relativity, but before Einstein's general relativity, gravity (which remained Galilean) and the rest of physics (which had become governed by special relativity) were in disagreement: here, again, it was to Einstein's credit to realize that it was not enough to "make relativistic" Newtonian gravity, but that a new framework was needed, implying still more general unique transformation laws. General relativity achieves a marvelous new unification of geometry and matter-energy, of gravitation and motion.

Has this story been finished? Is physics currently balanced and harmonious, ruled by a single system of laws founded on first principles? It is not, quite the contrary. The crisis of modern physics is at least as profound as that of Aristotelian science. Depending on whether one considers small or large scales (microscopic and macroscopic domains), the phenomena concerned, the experiments, or the observations that one makes, must be explained respectively by quantum or classical laws. Between quantum mechanics, which regulates molecular, atomic, nuclear systems as well as the physics of high-energy particles, and classical mechanics, adapted to ordinary scales, to those of planets, stars, our galaxy, and extragalactic systems, it is not only the laws which differ. It is a whole way of thinking, a manner of treating problems, the choice of mathematical tools and the rules which govern them, which differ at the most fundamental level. One could say, in a provocative manner, that physics currently is in a quasi-"schizophrenic" state, being not one, but two: two almost contradictory physics coexist in an anything-but-peaceful manner.

Thus the relativist "diagnosis" applies especially well with the contemporary physics. One will see that the principle of relativity allows us, once again, to propose remedies for this crisis.

Reference Systems

The second essential concept in the statement of the principle of relativity, after that of the "laws of nature," is that of a reference system. In physics, this more precisely means a "coordinate system." One appreciates the necessity of defining such a system in a clear and certain manner once one wants to make measurements. An essential characteristic of physical science (and what gives it its power) is that its statements concern results of measurement. However, even without measurement, none of our interactions with the external world would be made without the intervention of a reference system. It is what allows us to envision the universality of the principle of relativity and the possibility of its application to domains of knowledge other than physics alone.

The first and simplest measurement is that of an object's position. It is immediately clear that this concept has no meaning without reference to another body. Let us imagine a single isolated point within a completely empty space. There would be no means of observing its position, nor any way to define the meaning of this word. It is similar with the orientation of an axis. Two points allow us to define an axis, but its orientation in three-dimensional space can only be defined with respect to three other axes constructed beforehand. Let us now imagine a boat in an empty space, with no reference. How would we know whether it is moving or not? Here again, without reference, the notion even of movement loses all sense. Finally, let us consider the same boat, still in a vacuum. How can we measure its size, without relating it to another object that can act as unit of comparison? The scale of the boat has no meaning in itself, one must relate it to a body of reference.

Ultimately, a frame of reference can be defined as an abstract system which synthesizes the universal properties of the mechanism of measurement, that is to say, those which are independent of the specific characteristics of the instruments effectively used, but which are common to all these instruments. It is in this way that the result of measuring a point's position along one direction is given uniquely by a number, expressed using a certain unit, associated with an error bar, for example (12.35 ± 0.02) m. This result requires only the definition of an origin, an axis, and a unit which act as references, and a resolution for the uncertainty of the measurement (the resolution is determined by the minimal unit accessible by the given instrument, for example the interval between the two tick marks closest together on a ruler). However, to obtain results in practice, much more has been required than the quasi-abstract characteristics which are a perfect point of origin, an infinitely thin axis and markings traced on this axis. In reality, one must use a wooden, plastic, or other kind of ruler, and trace the markings with such and such a method and such and such a color. All these characteristics which are specific to the ruler actually used disappear as non-significant in the definition of the ruler seen as an archetype of an element of a coordinate system. It is similar with the details of the constitution of a clock: only the numerical value of the time that it gives is of importance, not its form or its inner composition.

Relativity of the Frame of Reference's State

We now arrive at the essential observation, what enables us to designate this type of theory as "relativity," even though it is in many respects a quest for an absolute.[61] But the absolute, which is translated in formal mathematical terms by the search for and definition of *invariants*, cannot be elucidated as long as the relative is not understood.[62] This is all the more true because the "absolute" is contained in the statement of the relative. Yet the great discovery of Galileo is that there does not exist motion in itself, only motion relative to another body, which he expressed in the following way:

> Among things which all share equally in any motion, it does not act, and is as if it did not exist.[63]

More generally, the physical quantities which characterize the state of the coordinate system (origin of the system, orientation of its axes, speed, etc.) can never be defined in an absolute manner, but can only be defined relative to another system. *They have no existence by themselves; only the relation between objects or systems taken two by two has meaning in physics.*

Here is an extremely profound statement on the ultimate nature of reality, which goes far beyond the science of physics. Indeed, the impossibility of defining a position, an angle, or a speed "in itself" does not correspond to a temporary limitation in instruments that an eventual future progress may resolve, nor to a provisional conceptual difficulty that new concepts will eliminate. It is the result of a much more profound and definitive fact: the position of an object, its orientation, its speed, or its acceleration—*these do not exist!* One can be convinced of this by imagining, as we have already done, a body which would be isolated in space in a universe which contained nothing else: the position of this body, its orientation, and its speed would be concepts void of meaning, impossible to define.

One should never even speak of the "speed of a body," but only of "the speed between two bodies" or "inter-speed": the words which we still use today have been forged in a non-relativist mental framework and contain in them the error and illusion which were present at their construction.

Where does this illusion come from? Why does it seem so natural to speak of our position or the speed of a vehicle in itself, when only relative speed has meaning? Why, even, did Greek science, which otherwise achieved leaps as prodigious as Euclidean geometry or number theory, go astray on this subject? Why did it take more than two thousand years for humanity, since the invention of science, to finally understand the laws of inertia, to discover relativity with Galileo, and to make it triumphant with Einstein?

The answer lies without a doubt in the particular position where we humans are, tied and restricted inside coordinate systems with extremely limited possibilities of transformations. One can attribute a large part of the history of humanity to the relentless effort that it exerts to get around these constraints: progress in the understanding of the laws of nature precedes and accompanies the technological progress which allows the achievement of feats that had sometimes been understood centuries before as thought experiments. Galileo had been, by means of thought, one of the first voyagers in space; Einstein, in his inner vision of the equivalence principle, had virtually experienced in 1907 the feeling of weightlessness that all astronauts discover when they see their pen float beside them in their space capsule. Heliocentrism had to await Copernicus to be established, while Aristarchus of Samos had proposed it one thousand seven hundred years before.

Is it not because movements from one place to another (changes in the origin of the system of reference) only rarely exceeded several thousand kilometers up until the end of the Middle Ages? With the first voyages to Asia, then the discovery of the Americas, soon followed by the first circumnavigations, the changes in accessible point of view became the size of the Earth; one discovers the sky of the southern hemisphere and its completely new stars; the historic coincidence between the great discovery of Copernicus and the epoch of great navigators is perhaps not an accident. Similarly, the laws of motion in special relativity have been difficult to discover because we are confined within frames of references of which the relative speeds with respect to each other remain tiny in relation to the speed of light, c. If it were not so, and if, like elementary particles, our natural speeds were comparable to c, these bizarre laws (which are those of rotations within a four-dimensional spacetime) would be intuitive for us, simply because they would be a part of our everyday experience, like they are for rotations in space.

The development and the ever-deeper understanding of relativity thus accompanies an authentic development in the size of state changes of possible reference systems, first by thought, then in practice. In the case of scale relativity, which we will speak of later, the microscope and then the particle accelerator (at small scales) and the telescope (at large scales) have permitted an opening toward domains normally inaccessible to human senses. In this new approach, one considers that the zoom of a microscope or of a telescope, which puts in effect a transformation between scales, can be identified with a new type of change of our reference system.

Relativity, Not Relativism

"Everything is relative!" How many times has one read or heard this phrase, supposed to sum up the essence of relativity? But nothing could be further from the spirit of relativity than this statement, with its too-great generality and imprecision destroying its meaning. Often used as a defense of confusion, this cliché tends to make one think that "one can be sure of nothing," that "one opinion is as valid as any other." No! All is not relative. On the contrary, certitudes can be achieved in physics, once the framework where they are stated has been well defined.

What the principle of relativity affirms is exactly not that "all" is relative, for one of the goals and the main tools of relativity is the discovery and definition of invariants, in particular, of physical quantities which *do not depend* on the choice of coordinate system.

More generally, relativity is expressed in the *covariant* mode of description, that is, when the form of equations does not change in a transformation of reference frame. There indeed exists an "absolute" in the theory of relativity, but this corresponds to a higher level of abstraction. It is found at the level of relations rather than objects, and appears even in the awareness of relativity.

What the principle of relativity affirms is that there exist certain special quantities, characterizing the state of the coordinate system, quantities which can *never* be defined in an absolute manner. In a yet more general sense, one can say that these quantities bring about the interface between ourselves and the outer world, upon which we would like to make measurements.

Finally, what this principle affirms is that the always-relative character of *these* quantities and none other is not just a profound and still revolutionary statement, but a means of allowing one to *construct* the equations of physics themselves. It is enough, to be convinced of this revolutionary character, to remark how, after over four hundred years, after and despite each of its successes, the principle of relativity finds itself each time attacked and rejected for decades and sometimes centuries. Thus, when Copernicus abandoned the absolute character of the Earth's center taken as origin of the world, it still needed to be defended by Galileo *one century later* and remained so revolutionary that he risked his life to protect it. It is indeed exactly the same principle of relativity of motion, *such as Galileo had stated three centuries earlier*, that constituted the start of the twentieth century's scientific revolution when it was reintroduced by Poincaré and Einstein (with new consequences)!

Relativity and Construction of Laws

The key idea discovered by Einstein is that the principle of relativity enables one to construct the fundamental laws of physics. Einstein was particularly conscious of this power of the relativist approach. In a letter to Pauli from 1948, he writes: "I am a fierce partisan not of differential equations, but of the principle of general relativity, whose heuristic force is indispensable to us."[64]

The principle of relativity is in fact not only an explanatory principle, but also and above all a constructive one. In other words, the laws of physics have no need to have been posed or written by some external principle in the world. It is from merely stating their existence and from the logical character of this existence that the form of the laws can even be deduced.

How does one proceed concretely? One must first of all translate the principle of relativity, which is applied to ideal laws, non-written, to a more concrete principle, which is applied to the explicit version that we have of these laws, that is, to the equations of physics. The universal validity of a law, in all reference systems, is translated by the invariance of the form of equations in the changes of systems. This is what Einstein called the *principle of covariance*.[65] To construct laws then becomes a mathematical problem. After having formally defined a general class of coordinate transformations, one applies these transformations one or several times, then one requires that the form of the laws, including the laws of transformation themselves, be conserved. This constraint limits the mathematically possible laws to those which are physically possible, that is, which satisfy the principle of relativity. From the mathematical point of view, this means requiring the formation of symmetries (such as invariance by translation, by rotation, under spatial and temporal reflections, etc.), and then establishing the general form of laws which satisfy them.

A double dynamic is thus seen in the development of physical theories governed by relativity: generalization of possible transformations, which will not work without a generalization of the geometric framework and even of the mental framework, then restriction of geometries to those which satisfy the principle of relativity.

From the Relative to the Absolute

Our everyday experience shows us that the apparent size of objects changes according to the angle from which we view them. If we turn a ruler in front of us, it will seem to dilate and contract, but we know very well that it is only an effect of perspective, and that its length has remained constant.[66] This illusion comes from our vision being only two-dimensional, while space is three-dimensional.

It is similar with special relativity. The contraction of lengths and the dilation of time predicted by Einstein's theory in the case of a relative motion are themselves nothing more than effects of four-dimensional spacetime being projected on the subspaces of three and one dimension, which are space and time. Here again, relativity enables us to define a generalized length, in four dimensions, which does not change depending on the state of motion of the observer. In Einstein's general relativity, this invariant becomes much more complicated, for it can depend on the position in space and the instant in time under consideration.

One of the principal goals of the theory of relativity is precisely to determine which quantities are invariant when the coordinate system changes. In this sense, relativity is indeed a search for the universal through an analysis of the relative. It means going beyond the appearance of measurements, of which the numerical results depend on the choice of the coordinate system, toward an intrinsic description of objects, independent of all reference points.

More generally, one seeks a unity of physical description, which can stay valid in passing from one point of reference to another. We have seen that this invariance in form of physical quantities and of equations is what Einstein called covariance. The idea is to obtain a general representation of laws where the form no longer depends on the frame of reference in which they are expressed. Nevertheless, once it is a matter of applying such general laws to a particular problem, the choice of a reference point is imposed, and the covariant relations will then be translated by numerical values depending on this choice.

Chapter 7
Toward a New Extension of Relativity

Relativity as Method and as Way of Thinking

Relativity is more than a principle; it is a method for investigation as well. Huygens was one of the first physicists to have explicitly used the relativist method to solve the problem of collision laws, when all other scientists stumbled on this question. This method is simple: instead of trying to describe the system under consideration (here, two balls of different masses striking each other and then separating with velocities to be determined) from the terrestrial reference point on which the balls roll, Huygens understood that other reference points were much better adapted to the task.[67] According to the question posed, it would be much better to locate oneself at the perspective of one of the balls, or rather, at a reference point corresponding to its center of gravity. It is much simpler and more effective to solve a problem with a reference point adapted to it, and then to make the change of coordinate systems which corresponds to the observer's point of reference, than it is to look immediately for a solution in the latter coordinate system. The ordinary frame of reference, in fact, is in all likelihood not at all adapted to the system, and adds a supplemental complication to the problem to be solved. It is precisely relativity, with its statement that all reference points are equally valid for the writing of physical laws, that allows this choice of the "good" reference point.

Moreover, as Bergson understood, Einstein brought about a new way of thinking. With general relativity, science answers, perhaps for the first time, not only a "how?" (to which it is generally considered that it must be limited) but in part a "why?"

One can recall in this regard the two fundamental questions asked by Leibniz:

> *Why is there something rather than nothing?* For nothing is simpler and easier than something. Further, suppose that things must exist, we must be able to give a reason *why they must exist so* and not otherwise.[68]

If physics can clearly say nothing about the first question,[69] from the fact of its existential character, it is well equipped for answering the second, of course not in a definitive way, but by bringing progressive elements of an answer. For this, the principle of relativity plays an essential role through its capacity not only to describe, but to explain. It can, in fact, be understood in a more general sense than its application to particular theories: we have shown that it is a general tool of work, a universal mode of thought, as well as a constructive method for obtaining the fundamental laws of physics.

It is true that the purely descriptive phase plays an irreplaceable role in the development of ideas in physics. But it is equally true that science has been able to arrive at an authentic understanding of certain phenomena (we have shown this, for example, in the case of the nature of gravitation), even if the answers given at a certain time, no matter their depth, always rest upon a deeper "why," which can remain long unanswered.

Let us note in this respect that the particular type of answer that physics provides for fundamental questions does not lead to emphasizing "causes." On the contrary, the question of "why" is found to be definitively circumvented when one shows the existence of certain phenomena as being imposed from the fact of the greatest generality possible. It is thus that Galileo discovered that uniform motion has no need of a cause; only rest or a change in movement requires one. A body resting on Earth seems to be self-evident, but this absence of relative motion indeed has a cause, and indicates the connection between the body and the Earth. If one considers two random bodies in the universe, the probability that they are at rest with respect to each other is practically zero! There is thus nothing to justify the existence of motion, which imposes itself (once the existence of space and time is admitted). Similarly, as we have recalled in the second part, Einstein's theory derives the existence of gravitation as a consequence of the non-absolute character of spacetime, that is, of the general relativity of motion. A four-dimensional spacetime without gravitation (that is to say, flat, and thus absolute) would be unthinkable.[70]

Furthermore, it is a problem of what we call "understanding" that lies at the heart of the relativist way of thinking. Has there not been a connection between relativist method and understanding, and thus, at a certain level, between the relativist mind and scientific creativity? The way which leads to understanding also advances with the possibility that one is ready to change reference point or system. To see things from many perspectives, to consider a problem in every sense, to invert, shake up, manipulate, and illuminate it from multiple angles, is often a key element of the discovery. In this process, a formerly hidden facet can appear, or rather, the same object, turned inside out like a glove, can suddenly take on a totally new appearance. To discover is often to see otherwise.

Here again, a particular point of view plays an essential role: that which consists of putting oneself "in the place" of the system under consideration, to envision it from the inside and not only from the outside. It is a matter of placing oneself in the frame of reference best adapted to an understanding of the problem, in a sort of "empathy" with nature. Such mental exertion is required for one to speak of true understanding: one knows because one sees and directly senses, even though virtually, the nature of the system to be understood. Such was the privileged method adopted by Einstein, when he anticipated special relativity in trying to imagine what an observer travelling with a light wave would see or, later, when he laid the foundations which would lead to general relativity by mentally visualizing the experience of free fall in a gravitational field.

Beyond Spacetime

Is it possible to go further? One can ask if all the possible theories of relativity have been constructed. Has the principle of relativity been pushed to its utmost limits? Have all its consequences been understood and described? We have seen how the development of theories of relativity advances by developing the definition of coordinate systems. Have all possible states of reference systems and all the transformations which tie them together been taken into consideration? It is possible that one has not yet fully analyzed the entirety of quantities which characterize our measurement results and that all invariants have not yet been discovered. Finally, is curved spacetime a large enough frame to understand all of physics?

It seems that Einstein did not think so since, starting from the development of the theory of general relativity in 1915-1917, he dedicated himself to the quest which he vainly pursued for the rest of his life: that of an even more extensive theory of relativity that would explain not only gravity, but the electromagnetic field and quantum effects.[71]

It seems legitimate to ask oneself anew today the question of a possible extension of relativity, without, of course, hoping to attain results as vast. In fact, our understanding of quantum field theory has made enormous progress in recent years, while new mathematical tools have come to light. In particular, new geometries of spacetime can now be imagined, such as fractal geometries, within which one can define transformations between coordinate systems hitherto unknown.[72] In the remainder of this book, we will see how one can apply the principle of relativity within the framework of a spacetime possessing such a fractal geometry, which leads to possibilities of a new understanding of quantum theory.

And further yet? Certain great philosophers have gone beyond, concluding the nonexistence in itself of all things, of matter as well as of mind. If we have been able to trace the history of relativity to Copernicus as far as occidental thought and the materialist sciences are concerned, its first statement in Asian thought seems to go all the way back to Siddhartha Gautama, more than two thousand five hundred years ago. One finds in Buddhist philosophy a truly relativist reflection on the emptiness of all things, consequence of their non-being in themselves, their existence only occurring in the relations between them.

One can only admire such an intuition, which we could consider an inner vision of the distant goal, perhaps inaccessible, that a science would propose based on the principle of relativity. Here there is no nihilism, nor negation of reality or of existence, but rather a profound view of the nature of existence itself. If things do not exist in an absolute manner, but exist nonetheless, their nature is to be found in the relations which unify them. Only relations between objects exist, not objects by themselves. Objects are thus empty in themselves (i.e., devoid of proper or inherent existence), and should be reduced to the entirety of their relations with the rest of the world. They are these relations.

Here one recognizes, made universal, the statements of the principle of relativity and of the equivalence principle, knowing their success in describing the nature of motion and of gravitation. In this case, Einsteinian physics has indeed put these ideas into effect: nonexistence of gravitation in itself, existence of gravitation as relative to the coordinate system, absence of gravitation in the system in free fall, inertial laws in this system, which are those of motion in a vacuum lacking all force, and so on.

Will the physics of the future succeed in putting into equations what currently amounts to a purely philosophic vision? This question obviously goes well beyond the limited scope of the present work.

Part Three
Quantum Mechanics

Due to the impossibility of explaining certain experimental results in microphysics using classical concepts, physicists created the field of quantum mechanics. In its initial version (that of the Copenhagen interpretation), it was a minimal theory, obtained by pruning away apparently unnecessary notions and piecing together those closest to the observed facts. The trajectories of particles are unobservable in the classical sense: this concept is removed. In numerous situations, a system no longer evolves in a strictly causal manner: starting from initial conditions that seem perfectly identical, results can vary considerably. It becomes impossible to predict the evolution of individual trajectories in such a system. Nevertheless, one observes in these cases that the relative rate of different possible results is perfectly stable, such that the probability of obtaining a certain result stays the same: the theory is thus probabilistic in essence.

The principal property upon which quantum theory is constructed is wave-particle duality. This discovery, which we owe to Einstein and de Broglie, is without a doubt one of the most beautiful unifications in the history of physics. In 1905, when light was considered by all physicists as wavelike, Einstein showed that the photoelectric effect can only be understood if there exist light particles, which were later called photons. Light seems to be wave and particle at the same time. During the same period, the particle theory of matter, with the discoveries of molecules, atoms, electrons, and then protons, was established. In 1923, Louis de Broglie "lifted a corner of the great veil," as Einstein wrote to Langevin concerning de Broglie's thesis, in proposing that matter also possesses wavelike properties. When experiments on electron diffraction verified this prediction, it allowed for attributing a similar status to light and to matter.

The basic tool of quantum theory simply and directly combines these three elements (probability, wave, particle) in a single theoretical object, the wave function, that is able to describe interferences (wavelike properties) due to its complex nature (in the sense of *complex numbers*, that one can think of as vectors on a plane: the simplest wave function is composed of two quantities instead of one). The square of the modulus of this wave function (square of the length of the vector which represents it) gives the *probability* of observing the *particle* in a given position, moment, and/or state. In this way, one of the principal mysteries of quantum behavior can be described (if not understood): it is the *wave functions* (that is, vectors) that are added together if an event can be brought about in several different ways, and not directly the probabilities as in the classical case. And as the sum of two non-zero vectors can be zero (if they are opposite), an event that can be brought about in two possible ways, each very probable in itself, might become impossible!

Schrödinger and Heisenberg succeeded in writing (in two different but equivalent forms) the equation governing the wave function, and were followed in more general cases by Pauli and Dirac. These equations, as well as the correspondence principle associating operators acting on the wave function with observable physical quantities (so-called observables), constitute, with several other rules, the axioms of quantum mechanics.

Chapter 8
Principal Axioms of Quantum Mechanics

In its current state, quantum mechanics is an essentially axiomatic theory. Its forms and methods were established as a minimal conceptual framework which could account for experiments carried out at the scale of atoms, nuclei, or elementary particles. Instead of deriving or deducing its fundamental equations, such as the Schrödinger equation, from first principles, one must accept them as axioms. Quantum mechanics is a set of rules which have become established through their success, but for which the true source remains unexplained.

Certain physicists have even seriously predicted that it will always be impossible to understand quantum mechanics, that it must always remain beyond human intuition. Our intuition is only applicable to scales where direct experimentation is possible. Other domains are definitively foreign to us. However, this is to forget that the scales of cosmology are just as far from our own as those of elementary particles, and that intuition, even in the classical domain, is not given by itself but is constructed and developed by experience and learning. If quantum mechanics has not been understood until now, it is not necessarily because it is definitively non-intuitive, but perhaps because a new type of intuition needs to be constructed. Along these lines, one can observe that what we call intuition is often geometric in nature (is this because sight is the most developed human sense?), and that the unintuitive character of quantum mechanics could thus be tied to its absence of spatiotemporal interpretation (or better yet, foundation).

Not being based on first principles the meaning of which would be prior to equations (which is the case in general relativity), quantum theory must be interpreted *a posteriori*. Its equations have first been written, and their meaning elucidated afterward! Schrödinger had originally believed that he had proven his equation in the article where he established it, at the start of 1926. But here is what he wrote of this "proof" in his second article:

> We have only briefly described this correspondence [between the mechanical equation and the wave equation] on its external analytical side by [a] transformation which is in itself unintelligible, and by the equally incomprehensible transition from the *equating to zero* of a certain expression to the postulation that the *space integral* of the said expression shall be *stationary*. (Note: This procedure will *not be pursued further* in the present paper. It was only intended to give a provisional, quick survey of the external connection between the wave equation and the Hamilton-Jacobi equation.)[73]

For several years, Schrödinger maintained hope that one could still arrive at a real (not complex) wave function. Other founders of quantum theory convinced him otherwise. Finally, in 1927 Max Born proposed what would become established as the definitive interpretation of the wave function, four years after the introduction of matter waves by de Broglie, and two years after the major conceptual leaps of Heisenberg and Schrödinger.[74]

We will not enter into all the subtleties of quantum theory here, but simply try to explain its principal axioms by illustrating them with some of the experiments which have played a central role in its construction and interpretation, first and foremost that of Young's double-slit experiment.

Probability and the Wave Function

The first axiom of quantum mechanics defines the new tool of this theory, the wave function. We must let go of the idea of the *a priori* well-determined variables of classical theory in favor of probabilistic description. This does not mean that certain variables cannot be determined perfectly well. We must first of all object to a "fuzzy" picture of the quantum universe, where nothing would be fixed, and uncertainty alone would prevail. Instead, quantum mechanics changes the mode of description. While classically one may speak of the position of a body, in quantum theory only the probability of the body being in such and such a position can be defined. However, the classical situation still fits within this descriptive mode: it would be the case where the probability of a body being in a given position is one, and thus zero at any other position.

Moreover, the probabilistic approach already exists in classical theory, as in, for example, statistical physics. What gives quantum mechanics its extraordinary character is that one no longer directly defines probabilities, but instead defines a "wave function," starting from which the probability is calculated. This function, also known as the "probability amplitude" when it is defined between two spatiotemporal events (following Feynman's terminology), carries more information than just the probability. In the simplest case, the probability amplitude is given by two numbers, only one of which defines the probability. But the essential point is that, while the reduction to a single number can be compared to experimental data, one must permanently "carry" the two together if the theory is to agree with the experiment. The mathematical entity that describes the state of the system thus contains a hidden part that we call the phase of the wave function. One can have an approximate idea of the nature of this phase by comparison with the description of classical waves: ocean waves are also characterized by two quantities, their height (amplitude) and the distance between two crests (the wavelength, which appears as the phase).

As Feynman explains in his book *QED: The Strange Theory of Light and Matter*, to go from ordinary probabilities to quantum probabilities means going from a number to a vector. Indeed, it is in their mode of combination that probability amplitudes differ the most from classical probabilities and allow for certain apparently extraordinary behaviors. The wave function can be thought of as a small arrow. This arrow (or vector) has a length and an orientation, defined by an angle in the simplest case (two dimensions, corresponding to a vector defined within a plane). The square of this length gives the probability and the angle, the phase.

The method of combining probability amplitudes with each other depends on the accessible information. If an event can come about in two different ways, but it is possible to know at any moment which one has been "chosen" by the system, the probabilities are combined classically: the global probability is the sum of probabilities of each path taken independently. But if it is impossible to know which path is followed, the resulting probability amplitude is the sum of the amplitudes of each path. This means that one adds vectors and not numbers: it is a matter of putting together two arrows. The final probability will be the length of the resulting arrow, squared (see Figure 2).

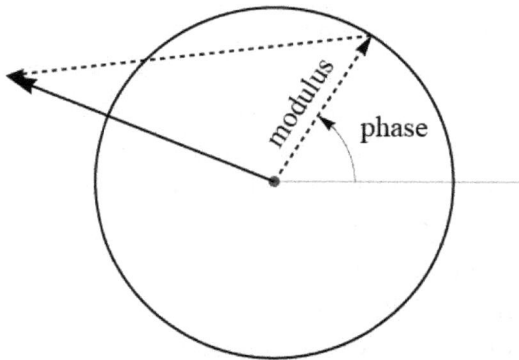

Figure 2 *Probability amplitude.* One calculates probability in quantum mechanics by squaring the length of a "probability amplitude" or "wave function." The amplitudes behave as vectors in an abstract space, in particular when one adds them together. The probability amplitude is defined not only by a modulus, but by a phase as well.

Here is what Feynman writes concerning quantum mechanics and its rules:

> One [has] to lose one's common sense in order to perceive what [is] happening at the atomic level. . . . In 1926, an "uncommon-sensy" theory was developed to explain the "new type of behavior" of electrons in matter.

> Will you *understand* what I'm going to tell you? . . . You think I'm going to explain it to you so you can understand it? No, you're not going to be able to understand it. Why? . . . That is because *I* don't understand it. Nobody does.

> The theory of quantum electrodynamics describes Nature as absurd from the point of view of common sense. And it agrees fully with experiment. So I hope you can accept Nature as She is—absurd.[75]

The result of combining quantum probabilities, which remains deeply incomprehensible but was established because it "works" (theory and experiment are in extremely precise agreement), is sometimes highly paradoxical. Indeed, the probability is given by the length of the arrow, but the sum of two vectors can be zero when neither of the two vectors are! For example, it is similar with velocities, which are vectorial quantities: if somebody walks on the deck of a ship in the opposite direction and with an equal speed, they stay immobile with respect to the dock, even though neither the speed of the boat with respect to the dock nor the person's speed with respect to the boat is zero.

But, if such a phenomenon is easily comprehensible for magnitudes which have a self-evidently vectorial character, this is certainly not the case for probabilities. Velocity can be positive or negative, but probabilities are always positive. Yet this rule of combining wave functions allows for the following incredible result: an event which can come about in two different ways, each highly probable, can become impossible (its probability is cancelled out).

Inversely, its probability can also become greater than the sum of the individual probabilities.

Young's double-slit experiment illustrates this property. Two slits made in a screen are illuminated by the same source. This source can be luminous, but, as Louis de Broglie was the first to understand, one can also perform the experiment using particles of matter such as electrons and neutrons, and even with atoms: these last two quantum objects, while complex and possessing an internal structure (neutrons are made up of quarks, while atoms are formed by a nucleus surrounded by an electron cloud), act as particles from the point of view of this experiment. Two extreme types of Young's double-slit experiment can now both be performed: one where the slits are lit up with a source of light or radiation (which corresponds to an enormous number of particles, eventually indeterminate), the other where particles (photons for light, or particles of matter) are sent one by one through the slits. The latter case highlights the entire quantum paradox.

The slits make the radiation diffract, so that, at its exit, it can travel in all directions. In fact, they simply act like two identical sources (which would be much more difficult to do if one did not start with a single unique source). The light rays (or the individual particles) are finally captured on another screen situated farther along.

The first part of the experiment consists of only keeping one of the two slits open. After having passed through the slit, the radiation strikes the second screen in a pattern forming a continuous distribution with a single maximum (this distribution characterizes those of the angles of diffraction by the slit). For the experiment performed particle by particle, we cannot precisely predict where any individual impact will be, but we observe that the distribution of impacts, gradually as their number increases, more closely approaches the distribution of intensity made by a strong ray. This fact fully justifies the interpretation of the square of the wave function's modulus as a distribution of probability density: in the case of a strong ray, the number of particles is such that this distribution is instantly realized, but the experiment of "filling it in" particle by particle explains its true nature.

Next, the second experiment involves leaving both of the slits open and comparing it to the result obtained in the preceding case. If the light ray or particles had been classical, one would expect to obtain a distribution given by the sum of distributions corresponding to each of the slits opened individually. But nothing of this kind happens. The distribution obtained, whether made immediately by a strong ray, or gradually as the sum of individual impacts, shows a pattern alternating between dark and light bands, that is, a distribution of probability showing successive peaks and valleys. In such a way, there exist certain points on the screen where a great number of particles might strike if one or the other slit is open, but where no particle ends up when both are open! How is this possible, when one has apparently increased the number of possibilities in opening a second slit? In the case of the experiment performed particle by particle, one thus sees each individual particle travelling one by one, systematically, to certain predictable zones known in advance (those which will become the bright bands after a great number of impacts) and avoiding others. This is one of the principal aspects of the quantum "mystery."

Certain commentators have not hesitated to write that everything happens as if each particle, passing through one of the holes (but which one is not known), was "informed" of the second opening. Others have suggested that the particle in fact passes through both holes at the same time. The first perspective privileges the particle-like aspect of the quantum entity, the other its wavelike aspect. A more complete picture consists of speaking of wave-particle duality, and considering, with Niels Bohr, that these two aspects cannot be simultaneously observed, following the theory's principle of complementarity.

One can object to all of these attempts at description that they remain ultimately attached to classical concepts and methods. Quantum mechanics perfectly explains Young's double-slit experiment, but it does not do so with the help of a wave in the classical sense of the term (which corresponds to the properties of a fluid like air or water), nor with the help of a classical particle (constructed on the model of a billiard ball). The quantum method involves a new concept, that of a *complex wave of probability*, which cannot be reduced to classical concepts, nor even to a combination of classical notions.

In fact, even wave-particle duality is not quite an appropriate concept, in my view, for the description of quantum effects. Wave-particle duality presupposes that the entity is either wave or particle according to the type of experiment. But what one really does in quantum mechanics, that is, what works to correctly explain and predict experimental results through theory, consists of permanently "carrying" a probability wave defined by a probability (square of the modulus of the vector describing this wave) and a phase (orientation of the vector).

All attempts at ignoring one of these two quantities when performing calculations will lead to erroneous results, in disagreement with experiment. What are the wavelike or particle-like behaviors under these conditions? Is the behavior with a single slit particle-like, and that with a double slit, showing interference patterns, wavelike? But in both cases, one observes the probability alone, never the phase! The phase of the wave function is never observable, except indirectly. What actually happens in the double-slit experiment is that, due to the nature of adding wave functions (vector quantities), the term of phase difference *in the single slit experiment* is recovered in the (observed) probability of the wave function in the double slit experiment. Thus, it is the wavelike nature of the single slit experiment which is behind the wavelike nature where the two slits are open. The phase of this second experiment would only be observable *a posteriori*, as part of a *new* experiment of interference.

What is there of the particle-like aspect? Can one abstract it away between measurements and only allow that it appears at the moment of measurement? Here again, such an interpretation does not seem to me to be appropriate for the quantum method. The equations of quantum mechanics describe the development over the course of time of a probability wave for an entity being at a given position or in a certain state. This probability would lose all meaning if there were no longer any particles: it would become devoid of significance.

Moreover, every time one tries to address this question experimentally, the answer is clear: the detectors show a well-localized distribution of energy, momentum, charge, or other physical quantity which attests to the existence of a particle. In such a way if one asks the question—which hole did the particle pass through?—one can try to respond experimentally by placing a detector immediately behind the two slits. It is even possible to detect the passage of the particle (via another quantum figure other than position or speed) without affecting its essential properties in any way from the point of view of the experiment in progress. The result is known: the entity is always detected behind one of the two holes, never both at the same time.

But then, as soon as there is detection, even if it is non-perturbing, there is also a disappearance of interferences! Feynman had even imagined an experiment where a detector situated behind a slit would function in a random manner. The result is that the subset of detected particles does not form any interference pattern, while all the particles that have escaped detection are distributed according to the light and dark bands. Experimental projects proposed by Englert, Scully, and Walther have demonstrated that ultimately, whether information concerning the path followed by the entity exists or not is sufficient in itself for deciding the result (interference pattern or not). This is in keeping with quantum mechanics, but not with numerous attempts at interpretation, in particular those which attributed the result of Young's double-slit experiment to Heisenberg's relation of uncertainty between positions and speeds. The moral of the story: every attempt at understanding quantum mechanics should be based on its mathematical tool (or, if possible, should make use of it) rather than contenting oneself by reproducing one of its interpretations.

The Correspondence Principle

Another axiom of quantum mechanics that makes its intuitive understanding even more difficult is the correspondence principle. As the formalism of the theory became more developed, its creators replaced the usual physical quantities of classical theory (the observable properties, or observables, such as position, momentum, energy, etc.) by operators, described in terms of complex numbers, which act upon the wave function (i.e., they transform it). The construction of these operators is done by rules of correspondence which were established by practice, but which have no character of universality or uniqueness. They can be reduced, by a sort of change of reference, to a set of numerical values which characterizes them (that we call the proper values of the operator), and which provide the possible values that the observable under consideration can take.

Schrödinger's Equation

The equation for the wave function's progress in time and space was established by Schrödinger and, in an equivalent form, by Heisenberg. This equation can be constructed by correspondence starting from the equivalent classical equation (which is the one giving the total energy as a function of kinetic and potential energies), but is not truly proven. It is thus generally considered as one of the axioms of quantum mechanics. Schrödinger himself, de Broglie, Klein, and Gordon generalized it for the relativistic case, and, for the important case of the electron (which possesses an original quantum property, spin), it was Dirac who generalized it.

Collapse of the Wave Function

One of the main axioms of quantum mechanics, essential for its interpretation and for the theory of measurement, is that of the collapse of the wave function. This axiom, formulated by John Von Neumann, states that immediately after a measurement, the system is in the state given by the measurement. It plays an essential role in the rationality of the theory and in its relation to classical theory. In its absence all derived interpretations would be possible, in particular that of a vision of the quantum domain as a realm of perpetual uncertainty.

Von Neumann's axiom ensures the continuity of the way in which quantum systems develop. For example, applied to positions, it excludes the possibility of a particle jumping from one point to another in space: immediately after a particle has been measured at a point, one is sure that it is in the neighborhood of this point. Applied to a polarization, this axiom assures the agreement with experiments made with polarizers: if we polarize light by making it go through a polarizer and no perturbation intervenes, we can verify with the help of another polarizer that the light has maintained its polarization. One speaks of wave function collapse since the measurement makes the entity go from a state described by a set of possible values of the observable to which various probabilities could be attributed, to a new state where one single value remains possible (the probability 1, which signifies certitude, is thus assigned to it).

Chapter 9
The Paradox of Quantum Properties

Einstein-de Broglie Relations

The association, discovered by Einstein and de Broglie, between, on the one hand, a free system's energy and momentum (the system can be either matter or radiation) and, on the other hand, a wavelength and a period (or inversely, a frequency) constitutes one of the principal mysteries of the quantum world. This structure is incorporated into quantum theory, but not truly understood. In this way, the Schrödinger equation can be constructed as the nonrelativistic equation of which such a wave would be a solution.

Energy-momentum constitutes, in classical mechanics, the most fundamental characteristic property—the most "primary"—of physical objects. It acts as "charge" for gravitation; everything that exists—from the point of view of physics—possesses energy-momentum. All other physical properties can disappear (there exist electrically charged bodies and others uncharged, bodies possessing or not possessing a strong or weak charge), but energy-momentum cannot be removed without the existence of the object itself being nullified. This universality of energy-momentum gives an idea of the extremely fundamental character of the Einstein-de Broglie relation. All physical objects possess *an energy, to which corresponds a period which is inversely proportional.* Similarly, if it is moving relative to a reference point, the object possesses a momentum (product of its mass by its velocity): *corresponding to it is a wavelength which is inversely proportional.* The constant that connects them is Planck's constant, h.

The existence of the de Broglie wave has been verified (by experiments using interference and diffraction) on all sorts of particles, whether coming from radiation (photons) or matter (neutrons, protons, electrons), but it also exists for complex objects like atoms or molecules. Why are energy and momentum thus universally associated with intervals of time and of length, those which characterize the de Broglie wave? This is one of the main questions which every attempt at understanding quantum phenomena starting from first principles ought to address.

Heisenberg Relations

One of the best known symbols of quantum theory is constituted by the Heisenberg relations that give rise to the so-called "uncertainty" principle. These relations do not belong to the axioms of quantum theory, as they can be mathematically derived starting from a general physical property which is not in fact specific to the quantum domain: that which defines how certain variables (such as position in space or moment in time) relate to the variables which are their conjugates (such as momentum or energy).[76]

From this point of view, quantum theory reinforces links which had already been established by classical theory. Let us recall how physical quantities which are conservative (invariable over the course of time) are naturally constructed based on the symmetry of certain variables (that is, on their invariance in certain transformations). It is thus with energy starting from the uniformity of time, with linear momentum starting from the uniformity of space, and angular momentum starting from its isotropy.

The Heisenberg relations of quantum mechanics are constructed using the same pairs of variables: time-energy, position-linear momentum, angle-angular momentum. They state that the product of uncertainties with which we can know each of the two variables of such a pair is always greater than a universal constant, equal to Planck's constant h divided by 4π.[77] They thus take the same form (in terms of inequality) as the Einstein-de Broglie relations, but dealing with dispersions and not averages. The consequence is that if one of the variables is known with greater precision (for example, the position), then the uncertainty of the other (here momentum, and thus speed) increases.

These relations have often been interpreted as a sort of loss for physics, a limitation sometimes deemed intolerable. The truth is that there also exist error bars and uncertainties in classical theory (and that one of the essential methods of physics consists of always taking them into consideration when interpreting results); more importantly, the precision of classical measurements is in general so relatively inexact, and the value of h so small that the product of the uncertainty of the position and velocity of a classical system is always much, much greater than the quantum limit. Thus, in actuality, the Heisenberg relations have taken nothing away from physics, but have on the contrary added a new constraint between variables which were not before connected. From this point of view, it is a matter of relations of *certainty* rather than uncertainty: while one could in no way speak, starting from classical mechanics, of the relationship between resolutions in position and velocity, one is now aware that, due to quantum mechanics, they are connected by the Heisenberg relation.

Spin

With spin, we arrive at the first quantum value *that has no classical counterpart*. It demonstrates the absolute originality of the quantum object.[78] However, it is also very difficult to picture the phenomenon.

The existence of spin, a kind of internal angular momentum of the electron and other particles, was first established with the goal of explaining experimental data. It was introduced by Uhlenbeck and Goudsmit to account for the observation of electron states which could not be understood using ordinary quantum numbers. The concept of spin was thus created.

But the inevitable advent of this new physical property marks a new step in the gap between classical and quantum mechanics. Indeed, while quantities such as energy, linear momentum, and angular momentum behave differently in the two mechanics, their classical existence is still certain. Spin simply does not exist classically. On the one hand, an electron might be completely pointlike, and thus not have any angular momentum.[79] It might, on the other hand, possess spatial extension, but one could then calculate its speed of rotation as a function of the measured value of its spin ($\hbar/2$), and one would find that its surface must rotate more quickly than the speed of light! This contradiction led to the conclusion that spin was a purely quantum value, without any classical analogue. Ultimately, it was shown to be the first of a long series of other quantum quantities (such as isospin, lepton number, baryon number) which had to be introduced to explain various observed properties of elementary particles.

Spin also plays an essential role since its existence brought renewed attention to the nature of what, from the perspective of classical theory, was considered to be radiation (or field of interaction) and what was considered to be matter. In fact, two types of quantum values for spin exist, which correspond to collective behaviors totally different, if not the opposite of each other. Certain particles, such as quarks, or the protons and neutrons of which they are the components, possess a half-integer spin.[80] This type of particle, known as a fermion (from the name of Enrico Fermi, who along with Paul Dirac established their statistical properties), is governed by the Pauli principle, according to which two such particles cannot be found together in the same state. The consequence is that they cannot be gathered together in large numbers (there cannot be many together at the same point) and are not able to act as the carriers of a field of interaction. Nevertheless, this property allows them to make stable structures, like bricks in construction. It is thus that the association of three quarks can form the proton and the neutron, which can join together in nuclei, which form atoms along with electrons. All known matter is composed of these elements.

The particles of integer spin (called bosons, from the name of Satyendra Nath Bose, who together with Einstein established their statistics) have the inverse property: "extroverted" characters, they tend to assemble in groups of which the number is in general not even known.[81] This property, shared by photons, the weak W and Z bosons, and gluons as well, allows them to be vectors of fundamental interactions (respectively electromagnetic, weak, and strong). However, bosons cannot form structures since such structures would collapse.

These results of quantum theory continued in a remarkable way the start of a unification achieved by Einstein and de Broglie. Not only do light and matter have particle-like and wavelike properties, but their description is now similar. A set of particles constitute a quantum field (a Dirac field for the electron, an electromagnetic field for the photon), characterized using the same method in all cases. The differences in behavior, as significant as they are, come only from the difference in spin! This fantastic success gave hope to physicists to go further yet, and to complete the unification in allowing for the existence of a transformation between fermions and bosons. These would then be no longer two different types of particles, but the same type of particle in different states. This postulated new symmetry between elementary particles, called supersymmetry, has not yet been verified by experiment.

Indistinguishability of Identical Particles

Indistinguishability is one of the surprising properties specific to quantum objects, and it also has no classical counterpart. It is connected to a profound change in the concept of identity in quantum mechanics, and is possible due to the quantization and discretization of the physical properties defining not only elementary particles, but their combination in more complex structures (nucleons, nuclei, atoms).[82] Thus, all electrons possess exactly the same mass, the same charge, and the same spin.[83] Their identity goes well beyond anything imaginable classically. As similar as two macroscopic objects can be, their strict, absolute identity is impossible. Two electrons, on the other hand, are identical to the point of being completely indistinguishable, not because the information which would allow us to distinguish between them is inaccessible to us, but because this information does not exist at all. This fact has practical consequences: the quantum object composed of two electrons can no longer be considered as the sum of two objects, each individual electron. Rather, it is a new object, with different statistics. The exchange of two electrons simply makes no sense physically, since no "label" would allow us to identify them. The system obtained after the exchange is rigorously the same, so thoroughly that the "two" systems should be considered as one from the statistical point of view instead of two, classically speaking. This has been confirmed by experiment.

Inseparability, Entanglement, and Nonlocality

The Einstein-Podolsky-Rosen (EPR) paradox, in its modern version of John Bell's inequalities and the experiment of Alain Aspect, highlights in the most striking manner the difference between classical and quantum mechanics.

Two particles having spins which are the inverse of each other, but not defined, are emitted in two opposite directions. The spin values are then measured at the same time for the two particles. The quantum information arises here solely from an anticorrelation between the particles. Neither of the two spins have a well-defined value, but their sum is nevertheless zero with certainty. It is thus impossible to predict the value which will be measured for either of the particles (for example + ½ or - ½ with the same probability), but one can still be certain that if one measures the spin of one at + ½, the spin of the other particle must always be - ½, and it can thus be predicted with certainty starting from the first measurement.

Such a result might not seem very unusual from the perspective of classical mechanics. In classical mechanics, a measurement reveals the value of a quantity that preexists the measurement (which was not known to us due to lack of information about the system). It is totally different with quantum mechanics: the information known about the system is complete. No value for the spin exists before measurement. The measured value appears at the moment of measuring, but does not preexist it, like in an experiment of head or tails.

The physicist John Bell was able to show that the difference between these two situations—unknown hidden, but preexisting parameter, or complete absence of such a parameter—can be shown through experiment, since certain inequalities, strictly valid if there exist hidden parameters, should be able to be violated in quantum mechanics. The EPR-type experiment, performed by Aspect, has effectively demonstrated the definitive nonexistence of hidden parameters by verifying a violation of Bell's inequalities.

The EPR experiment becomes astonishing when one considers that at the moment when the spins of these two particles are measured, these particles can be arbitrarily distant; if one wants to reason in terms of transfer of information (in supposing that, as soon as one of the two spins is measured, the information about the result is transferred to the other particle, thus allowing the inverse spin to be obtained), this information must move faster than the speed of light, in contradiction with special relativity! It is as if one tossed two coins, one on the Earth, the other on the Moon (at a distance of about a light-second), at the same instant (with the assistance of clocks synchronized beforehand) and always obtained heads for one and tails for the other.[84]

But all attempts at using an EPR experiment to effectively transmit information are doomed to fail. This is why this type of interpretation has been rejected by most physicists. What quantum description expresses is that the "two" particles in fact constitute one single, entangled system: they are inseparable. This non-separability, as we have seen, in no way depends on the extent of the system, which necessitates another essential property of quantum objects (even "elementary"): their fundamental nonlocality.

These are some of the strange properties that any attempt at a new theory wishing to base quantum theory on first principles must be able to take into account. It now seems clear, after almost one hundred years of theoretical and experimental developments in the quantum domain, that attempts to return to determinism or the introduction of hidden parameters should be abandoned. Quantum mechanics, with all its paradoxes, must be included in any eventual future theory, in the same way that Newtonian theory is fully encompassed by Einstein's theory.

Chapter 10
Einstein and Quantum Theory

Einstein's position in relation to quantum theory is more subtle and complex than what tends to be said, which often amounts to caricature. No month goes by without some theoretical or experimental result confirming the predictions of quantum mechanics being the basis for headlines such as "Einstein was wrong..." However, to suppose that Einstein rejected these predictions constitutes a thorough betrayal of his thinking, which has only been denounced by some few commentators.[85] Einstein's thoughts are in fact clearly expressed in numerous texts on the subject.

Between the construction of the theory in 1925-1927 and around 1933, there is no doubt that Einstein had attacked quantum nondeterminism, especially regarding the Heisenberg relations. During these years, much of his efforts tended to think of experiments which would disprove these relations of uncertainty. But he was finally unambiguously persuaded by Bohr's arguments, since his critiques afterward gradually became aimed at a much more foundational level, that of the fundamental concepts of the theory.

Beginning in 1933, in fact, Einstein no longer argued that one must retract the nondeterminism of microphysics which quantum theory emphasized. In a lecture given around this time, he says:

It seems to me certain that we have to give up the notion of an absolute localization of the particles in a theoretical model. This seems to me to be the correct theoretical interpretation of Heisenberg's indeterminacy relation.[86]

It seems that he kept an open mind toward all possibilities. In a 1950 letter to Schrödinger, he writes:

The fundamentally statistical character of the theory is simply a consequence of the incompleteness of the description. This says nothing about the deterministic character of the theory.[87]

If, to the great despair of his friend Max Born especially (see the Einstein-Born correspondence), he strove to argue that quantum mechanics was an incomplete theory of reality, one must recall that he thought the same thing for the theory of relativity, of which even its "general" version he considered to be provisory, and indeed for all of physics:

Not for a moment, of course, did I doubt that this formulation was merely a makeshift in order to give the general principle of relativity a preliminary closed expression. For it was essentially not anything more than a theory of the gravitational field, which was somewhat artificially isolated from a total field of as yet unknown structure.[88]

One must still remember that for Einstein, there was no question that any future theory could be obtained by extending quantum mechanics:

I believe, however, that this theory offers no useful point of departure for future development.[89]

Einstein in no way went in the direction of theories of hidden parameters that he in fact never supported, contrary to what is often believed. He thus writes, in 1954, to Bohm, one of the principle physicists who had tried to develop such an approach:

> In the last few years several attempts have been made to complete quantum mechanics as you have also attempted. But it seems to me that we are still quite remote from a satisfactory solution to the problem.[90]

He expresses himself still more clearly on the interpretation of quantum mechanics in a text of 1953:

> The acceptable interpretation of the Schrödinger equation is the statistical interpretation given by Born.[91]

What Einstein could not accept as definitive in the theory of microphysics was that quantum mechanics was *by nature* statistical, that probabilities should be placed at the level of the theory's principles. This would abandon an explanation of individual elementary phenomena. Einstein, on the other hand, thought that any physical theory worthy of the name could not be probabilistic in its fundamentals, and that the necessity of a statistical description must, to be fully comprehended, be *derived* from more fundamental principles applicable to the description of "the state of individual systems." As we shall see in what follows, the theory of scale relativity comes under such a view: its fundamental foundation is in terms of nondifferentiable and fractal spacetime, and the statistical description is a consequence of the fact that geodesics are themselves fractal and infinite in number in such a spacetime.

> The argument (in quantum mechanics) completely leaves the processes affecting individual systems completely in the dark; these are totally eliminated from the representation provided by the method of statistical explanation.[92]

The physicist . . . will have to give up his position that the Ψ-function constitutes a complete description of a real factual situation. . . . The statistical character of the present theory would then have to be a necessary consequence of the incompleteness of the description of the systems in quantum mechanics, and there would no longer exist any ground for the supposition that a future basis of physics must be based upon statistics.[93]

If one wants to consider the quantum theory as final (in principle), then one must believe that a more complete description would be useless because there would be no laws for it. If that were so then physics could only claim the interest of shopkeepers and engineers; the whole thing would be a wretched bungle.[94]

It is this, and nothing else, that he meant by his so-often-repeated expression "God does not play with dice" (where "God" is to be understood in the sense of "laws of nature," Einstein often clarified), and in any case certainly not a rejection of the Heisenberg uncertainty principle nor a non-recognition of the gains and extraordinary successes of quantum mechanics:

[Quantum theory] represents an important, in a certain sense even final, advance in physical knowledge.[95]

It seems that what Einstein had in mind is that quantum theory must one day be encompassed by a larger theory based on different principles, in the same sense that Newtonian physics is encompassed by his theory of general relativity. In this process, the nonphysical concepts of the theory, such as that of a force acting instantaneously at a distance, disappear from the foundations (but might reappear as practical approximations). This is the fate which would thus be reserved, in this eventual future theory, for concepts such as the wave function and wave-function collapse.

Let us compare quantum mechanics and relativity. General relativity, as we have seen, is a theory that, even if it can become complicated in its application, is based on simple and comprehensible principles. On the contrary, the foundations of quantum theory are purely axiomatic in nature. Relativity is an essentially geometric theory, based on the primacy of the four-dimensional spacetime continuum (which even becomes its primary instrument). Quantum theory is, on the other hand, an algebraic theory for which the principal framework is constituted by spaces of abstract states: there often follows a loss of intuitive comprehension of phenomena. Finally, one can consider that general relativity contains conceptual levels deeper than quantum theory. In effect, the fundamental equations of Einstein's theory are those of the geometric structure of spacetime; the equation of trajectories (that is, of geodesics) is *deduced* from it; ultimately the equation of a *set* of geodesics is itself constructed starting from this latter equation. One might compare this last type of equation to Schrödinger's equation. Currently, there is no equation of an eventual "quantum spacetime." Here again, one can in this respect compare quantum theory to Newtonian theory in its time: an extremely precise theory, which "sticks to the facts," but remains unsatisfactory from the point of view of fundamental concepts.

In a letter to Louis de Broglie written at the end of his life, Einstein in part explained this point of view in specifying that the future theory should be a field theory (in the same sense as general relativity, that is, a theory of spacetime). Here, he is not alluding to a quantum field theory, which currently exists for fields such as electromagnetism and weak and strong interaction, but of a theory explaining quantum effects as the manifestation of a new field (ultimately meaning a geometry of spacetime).

In this sense, one must admit that none of the theoretical and experimental developments of the past few decades, as spectacular as they are, address in any way the problem posed by Einstein (and thus have brought no solution either way), since they are all situated (except the attempts at theories of hidden parameters which reject nondeterminism and are destined to failure after Bell's theorem) within a framework where the fundamental laws themselves are in essence probabilistic.

Regarding theories of quantum gravity, Jean-Pierre Luminet writes: "Hawking has deemed that the uncertainty principle applied to the black hole was transcended by what he calls the 'principle of randomness.' . . . Hawking's answer is the following: 'Not only does God play with dice, but He throws them where they can't be seen.'"[96] These theories are obtained starting from the *a priori* precept of quantization and the methods of quantum mechanics, and thus unfortunately do not address this fundamental question: that of the search for first principles which would be able to act as the foundation of quantum theory.

Part Four
**Scale Relativity,
Fractal Spacetime,
and Quantum Mechanics**

Chapter 11
Scales in Nature

Since antiquity, since the age of Plato, Euclid, and Aristotle, numerous philosophers and writers (such as Voltaire with *Micromegas* or Swift with *Gulliver's Travels*), in addition to mathematicians and physicists (among others, Leibniz, Boscovich, Laplace, and Poincaré), have considered the question of scales of length in Nature. If the philosopher or novelist has often imagined that there might exist men of lilliputian size or giants as large as the Earth, we know that in reality, one sees nothing of the sort. The height of an adult human can hardly vary by more than a factor of two, no one has a height of a millimeter or a kilometer. The size of atoms is of the order of a few angstroms, and there do not exist atoms with a radius of one meter, nor of one fermi: it is their nuclei which one encounters at the latter scale.[97] Inversely, the radius of a star is of the order of millions of kilometers and can reach one hundred times this size, but not one million times. At the scale of 10 kpc (ten thousand parsecs, the parsec being the astronomic base unit), the structures encountered are galaxies, but no galaxy is a thousand times smaller or bigger.[98]

Thus, any given structure is often characterized by a scale or a limited range of scales and, inversely, at any given scale there generally corresponds a certain type of structure. Why is it thus? What determines these scales? This is one of the essential problems, in large part unresolved, of fundamental physics.

The question is so much the more difficult since it is not *always* the case. Numerous physical systems, on the contrary, show no particular scale over a large interval (we call them "scale invariant"). Recall, for instance, the televised images of the first Moon landing. The observation, over the course of the descent of the lunar module, of its surface constellated by craters did not allow at any moment an estimation of the distance of the vessel to the Moon, due to the similarity of the craters to each other. In the same manner, estimating distances and depths becomes impossible on certain slopes on which rest rocks of all sizes. A small pebble up close becomes indistinguishable from a distant boulder without the help of another object for reference, such as a person or a house. Concerning the laws of gravity, Laplace writes: "One of the remarkable properties of Newtonian attraction is that, if the dimensions of all bodies in the universe, their mutual distances and their velocities grew or shrank in proportion, they would follow curves entirely similar to those which they followed before."[99] The substructures of a fern leaf are similar, within their scale, to the entire leaf.

Examples of scale-invariant objects and systems abound; a large number of them have been catalogued and studied by Benoit Mandelbrot, who came up with the word "fractal" to name sets showing the existence of structures at all scales (or over a wide range). But where do fractals come from? What are the physical principles which underlie their emergence? Why and how can there be a coexistence of scale-invariant laws or systems over a large interval of possible scales, and characteristic scales which break this invariance (for example, in the case of the fern, the size of its leaf at large scale and that of its smallest dendrites at small scale)? These questions have taken on a growing importance in contemporary physics.

There also exist objects whose very nature is to change scale. Growth and regression are characteristics of living organisms. The universe itself dilates over the course of time, and in its past went through a primordial phase over the course of which its characteristic size was enormously smaller than it is now, the phase described by big bang theory.

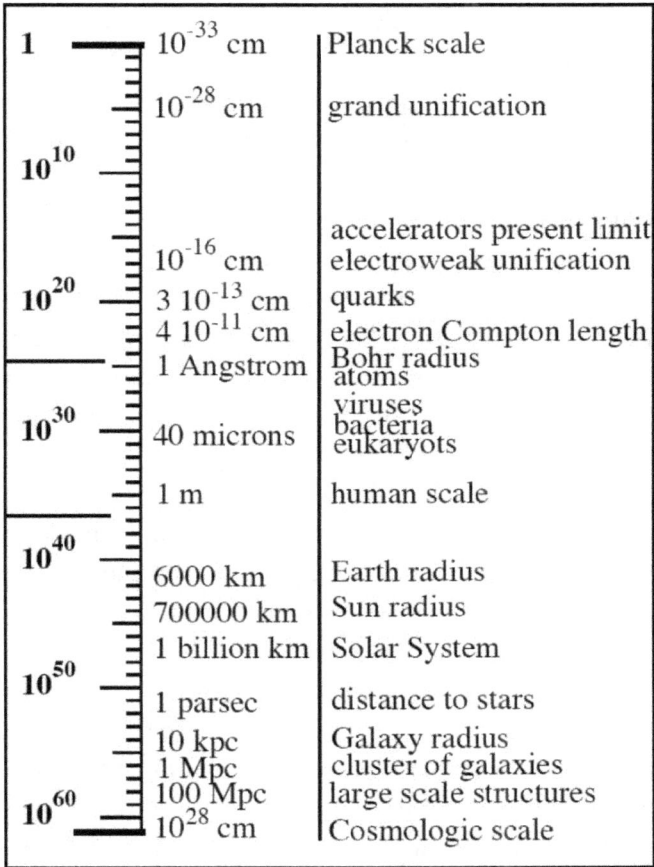

Figure 3 *"Scale of scales" in nature.* The distance between a tick and the next corresponds to an enlargement by a factor of 10. From the Planck scale to the scale of the cosmological constant the ratio is over 10^{60}.

Modern physics has certainly made enormous progress. But the origin of the characteristic principal scales remains insufficiently explained. It is thus that the characteristic size of atoms can be theoretically calculated by quantum mechanics and that of stars by astrophysics. However, this theoretical prediction involves the mass and charge of the electron (in addition to fundamental constants such as the speed of light c, and Planck's constant $\hbar$).[100] Similarly, the characteristic radius of stars can be calculated with the theory of stellar equilibrium, but as a function of the same constants, to which is added the gravitational constant G and the mass of the proton. Yet there currently exists no theoretical prediction of the mass of elementary particles (neither of electrons nor of quarks which form the proton). The values of their different charges (electrical, but also those which characterize their coupling due to other forces of fundamental interaction) are equally unexplained, essentially. The mass and charge of elementary particles are known to us only through their experimental measurements.[101] All of our theoretical calculations ultimately boil down to these measurements: for example, the set of energy levels of atoms, upon which rests all of chemistry, depends on the mass and charge of the electron and Planck's constant. The whole house of physics seems to be built upon sand.

At a certain level, we can say that not only do we not know how to calculate the value of the electric charge, but also that we do not really know what it is. The particular values of mass, in addition to their origins, remains in large part a mystery.

The American physicist Richard Feynman writes along these lines:

> There is no theory that adequately explains these numbers. We use the numbers all the time in our theories, but we don't understand them—what they are, or where they come from. I believe that from a fundamental point of view, this is a very interesting and serious problem.[102]

Yet this problem can ultimately reduce to a question of scale of length or of a ratio of scales. Quantum mechanics shows a correspondence between every mass and a characteristic length scale which is inversely proportional, called the Compton wavelength.[103] Theoretically predicting the value of the Compton scale of the electron leads to a prediction of its mass. Similarly, the ratio of this scale to that of Bohr (which limits the size of atoms) yields the value of the electric charge (See figure 3).

But the theoretical calculation of the atomic radius or that of stars does not only depend on the physical characteristics of their elementary constituents, which are electrons and protons. It also involves the numerical value of the fundamental constants G, $\hbar$, and c. The relation which ties the mass of the electron to its Compton wavelength also depends on $\hbar$ and c. These constants appear at all levels of the laws of physics. Certainly, a constant like the mass of the electron takes on a universal character, from the fact of the universality of the electron itself (it has been verified, through the study of the spectra of very distant extragalactic objects, that the electron was indeed the same as here and now, at a distance of gigaparsecs and a time of billions of years). But the universality of G, $\hbar$, and c is of another order. These constants are at play at the level of the fundamental laws of physics and do not depend on any particular object. Let us briefly recall their origin.

The gravitational constant, G, was initially introduced within the framework of Newton's universal gravitation. Its appearance is especially important for questions of units of measure. It is connected to the equality of inertial mass and gravitational mass which was discovered by Galileo and which forms the observational basis for Einstein's equivalence principle underlying his theory of general relativity. Newton introduced these two types of mass in the expression of two different kinds of forces. The fundamental law of dynamics states that the result of a force is an acceleration that increases as the mass of the body subject to the force diminishes. In other words, a force is equal to the product of its inertial mass and its acceleration.[104] Along these lines, gravitational force is proportional to the product of the gravitational masses of the bodies which attract one another and inversely proportional to the square of their distance. Equating the formulae of dynamics and of gravitational force thus necessitates the introduction of a constant, G, whose dimensions ($M^{-1}L^3T^{-2}$) (a volume divided by a mass and by the square of a length of time) will allow gravitational force to have the right dimension. But this constant acquires, in general relativity, a new and yet more universal status. It is what relates the geometry of the universe to its material and energetic content in Einstein's equations. It thus corresponds to something more fundamental than a simple constant characteristic of a particular field, in being connected with the geometry of spacetime. This constant is unfortunately one of the least well known in physics[105] (this may be linked to the fact that, in practice, it does not intervene alone but rather through its product by a mass: for example the product GM for the Sun mass is known with great precision).

The speed of light, c, also plays a role in assuring the homogeneity of physical units. It appears in the equations of physics only because one has chosen *a priori* different units for the measurement of length and of time. Nevertheless, since the advent of special relativity (within which it acts as a maximum speed, unsurpassable) and the understanding that neither space nor time have an independent existence, but are only the subsets of a four-dimensional spacetime, its status has become better understood. Similar to how we do not use different units to measure a length and a height in space, we should use the same unit for measurements of length and of time. In these conditions, speed is a truly dimensionless physical quantity, and the speed of light a fundamental "1." The question of its numerical value no longer has any meaning in itself, for it only reflects our inadequate choice of different units (the meter and the second). These theoretical findings have now been taken into consideration in the definition of units: since 1985, there no longer exist independent units of length and of time, but one unit only, that of time, of which that of length is deduced *via* the speed of light, now fixed in an *exact* manner.[106] This choice aims to maintain continuity with the previous choices: one might have, in a more radical (but equivalent) manner set c to 1 (without dimension) and measured lengths in nanoseconds, for example.[107]

Finally, Planck's constant is the fundamental constant of quantum mechanics. It was initially introduced in 1900 by Planck, in order to understand the nature of the law of blackbody radiation that he had just discovered. For this purpose, he made the hypothesis that radiation is not exchanged in a continuous manner, but by "quanta" of discontinuous energies. It was in 1905 that the theory of quanta was truly born, when Einstein made a giant step in postulating, in order to explain the photoelectric effect (in which electrons are removed from a metal by light), that the quanta were not only a property of the exchange of energy, but of light itself. Thus the light particle, which was later called the "photon," was discovered, with its dual nature of wave and particle. It is precisely Planck's constant, h, which connects these wavelike properties to the particle-like ones, that is, which connects the frequency of radiation v (which defines the color of visible light) and its energy E, following the Planck-Einstein formula $E = hv$. As we have already noted, this unification was finally completed twenty years later, when Louis de Broglie applied the inverse reasoning to matter: just as Einstein had shed light on the particle-like nature of radiation (considered as purely wavelike since the work of Fresnel and Young at the start of the nineteenth century), de Broglie postulated in 1923 the wavelike nature of particles like protons or electrons. This remarkable vision was quickly confirmed experimentally by the observation of effects of diffraction for electrons, and more recently on entire atoms: everything shows the universality of the Einstein-de Broglie relations. Planck's constant, which connects energy and frequency, as well as linear momentum and wavelength, possesses a dimension (that of angular momentum, ML^2T^{-1}, product of a mass by the square of a length divided by a time) that is fixed, here again, by the dimensionality of the physical quantities which it relates.[108]

Can we imagine one day predicting the numerical value of these fundamental constants by theory alone? Contrary to the mass or the charge of the electron, they are independent of any particular object. Their values depend on the units of physics, which have been chosen in an arbitrary manner. This problem is fundamentally a question of scale. In fact, starting from G, $\hbar$, and c, one can come up with three fundamental scales, a length, a time, and a mass, called Planck's scales. All laws of physics can be re-expressed in terms of these three new constants. In such a formulation, lengths and times operate only in relation to the Planck length and Planck time, and masses by their relation to the Planck mass. The equations of physics then become dimensionless. Would these three scales of Planck be natural units, allowing us to remove the arbitrary character of the choice of our units? Do we not thus lack a fundamental theory in which these magnitudes would naturally occupy this place, in the same way that, since the special relativity of Poincaré and Einstein, the speed of light plays the role of a natural unit for speeds? I will suggest answers to these questions in the remainder of the present work.

Let us return to the question of scale. One does not only see the existence of fundamental scales in nature, but also the existence of laws that are explicitly dependent on scale. In such a case, the physical quantities under consideration take on values which change according to the scale with which one measures them. To give a familiar example, it is so with the length of the coast of Britain, which clearly depends on the scale of the map which one uses to determine it. The more detail this map allows one to see, the more there appear new gulfs, inlets, and recesses which increase this length. Yet this variation of the length as dependent on the resolution (which is the size of the smallest observable detail on the map) is not made in any which way, but following a well-determined law. Such phenomena have led to concepts of scale laws and scale invariance, and to geometric description in terms of fractal objects.[109]

The dependence on scale can in certain cases be of an extremely fundamental nature. Thus, in quantum mechanics, the results of measurement explicitly depend on the resolution of the device which was used in the experiment, that is, of the smallest interval of the measured quantity observable with the device. Hence, to change the resolution in general requires one to change the measuring device. In cosmology, it is the scale of distances between galaxies which changes over the course of time due to the expansion of the universe: space in its entirety dilates. Moreover, one encounters scale laws and scaling behaviors in numerous situations, from small scales (in microphysics), to large scales (extragalactic astrophysics and cosmology), but also on intermediate scales (particularly in the domain of living beings). Most of the time, the discovery of such laws is made in an empirical manner. But it is rare that they are truly understood. We still lack a theory which would allow us to deduce them from fundamental principles.

The problems attached to scale do not stop there. One of these problems is one of the greatest puzzles encountered over the course of the history of science: this is the *appearance of quantum phenomena* in microphysics experiments. Not only is the nature of observed structures different according to the scale, but, as we have seen in the preceding chapters, the laws of physics themselves change between small and large scales. According to whether one studies microscopic domains (nuclear, atomic, and particle physics) or macroscopic (celestial mechanics, cosmology), the physical theory to use must pass from quantum mechanics to classical mechanics. Nor can we speak of an absolute scale of transition between the quantum and the classical: this is shown, for instance, by the existence of quantum effects in the macroscopic domain, such as in superconductivity (which allows, along with other effects, the transmission of electricity without loss). Inversely, in certain conditions, particles can be treated in a classical manner (it is thus, for example, with the image of their trajectory produced by a bubble chamber). Certain astrophysical problems at a large scale, like the internal structure of a neutron star or that of the very early universe, which both call for a description in terms of elementary particles, must make use of quantum mechanics. Nevertheless, there do exist relative scales of transition between the quantum and the classical domain, since, for a given system, the theoretical analysis should be classical at larger scales and quantum at smaller ones. It is simply that the scale of transition between the two behaviors depends on the considered system (on its mass, its speed, or its temperature), and is therefore not absolute, but relative.

This dichotomy in contemporary physics is extremely deep and troubling. It goes further than a simple change of laws in the classical sense of the term, which could have been attributed to the existence of new fields at small scale in relation to those known classically (gravitation and electromagnetism). Such new fields indeed exist (the strong and weak nuclear interactions), but the problem is not that. Everything changes between quantum and classical physics: the concepts, the mathematical tools, the manner of posing problems, even of thinking about them, the manner in which theory is formed, the manner in which it is understood.

It seems more and more clear, after a century of concerted effort concerning the relations between classical and quantum physics, that one cannot pass directly from one domain to the other: neither of the two is deduced from the other; they must coexist and be given independently. But is there no other way of unifying them? Cannot each of these two representations of the world be deduced from a third, more general representation, which encompasses them, and is reduced, according to the scale of the observation or the description, respectively to quantum and classical representations? It is such an approach which will be in question in the rest of the present work.

The idea serving as Ariadne's thread for this point of view is that there does exist a fundamental principle upon which to base the new theory: this is the *principle of relativity* itself. However, one must go beyond the principle of relativity of motion, as it has been developed by Galileo, Poincaré, and Einstein. It is necessary to introduce an extension of this principle which also encompasses the transformations between scales.

What can we expect from such an extension of our frame of thought?

As we will see, one first consequence is that it implies a profound change to the nature of spacetime. *The general relativity of Einstein had led to a generalization from the flat and absolute space of Newtonian theory to a curved spacetime, dependent on its material and energetic content. Similarly, the idea of scale relativity introduces a new spatiotemporal geometry, more complicated yet: spacetime becomes fractal.* The concept of fractals was expressly constructed by the mathematician Benoit Mandelbrot to designate objects, sets, or functions whose form is extremely irregular and fragmented at every scale, and for which the geometry thus explicitly depends on the resolution used in considering them (see figure 4 and the following). A new usage should nevertheless be made here. One no longer attempts to describe fractal *objects*, *a priori* included as contents of a space defined beforehand, but fractal *spaces* (more generally, fractal spacetimes), which should therefore be defined intrinsically, from the inside, since it is the container instead of the content. Moreover, as it is already the case in Einstein's theory of general relativity, it is not a question here of an absolute spacetime, but on the contrary, of the construction of a fractal spacetime that must be relative to its material and energetic content.

A second consequence is *a new possibility of understanding quantum mechanics*. We will see how elementary particles can be reinterpreted as families of "geodesics" (being the shortest trajectories from the viewpoint of proper time) of a fractal spacetime showing structures at all scales toward microscopic scales. More generally, the main axioms of quantum mechanics (meaning its mathematical principles, currently arbitrarily posed, and not understood) can be reconstructed and derived in the new approach starting from the principle of relativity itself (applied to motion *and scales*). Similar to how, in Einstein's general relativity, gravitation is nothing other than the set of manifestations of the curvature of spacetime, the theory of scale relativity suggests that quantum effects appear as manifestations of the fractal character of spacetime at small scales.

But we can go further yet. Scale relativity not only allows us to propose a source for quantum mechanics, it also leads to possible generalizations at very small scales. The application of the principle of relativity to scale transformations of coordinate systems leads us to propose possible forms for the laws which govern *the internal structures of spacetime*, structures we postulate to exist in each "point." Original fractal structures, more general than "ordinary" fractals, thus emerge. In this enlarged paradigm, *the problem of the infinite divisibility of space and of time finds a new solution*. It remains possible to divide a spatial or temporal interval in two, and then to divide that new interval in two, and so forth to infinity, but the result of this operation is no longer zero, but a finite interval. Similar to how one can add velocities to each other without end in special relativity, but the result of this sum always remains lower than the speed of light, here there *appears a universal, minimal scale, unsurpassable, upon which any magnification would have no effect.*

As we will see, it seems natural to identify this scale with the Planck scale, whose role as the natural unit of length and of time would find itself to be thus justified. This scale would then possess all the physical properties which were formerly attributed to the point zero and would replace it, all scales between zero and the Planck scale no longer having any existence (similar to how the speed of light possesses, in special relativity of motion, all the physical properties of an infinite speed).

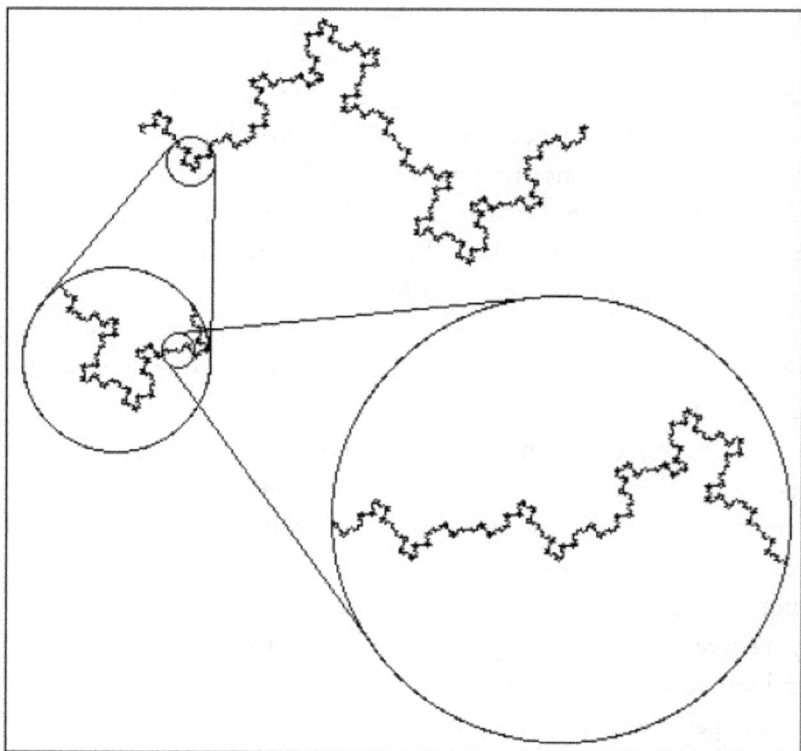

Figure 4 *Fractal curve.* A fractal curve shows new structures at all scales (in the particular case of this figure, which represents a self-similar fractal, it is the same structure that one finds at different scales). Such a curve depends explicitly on the resolution, which is the smallest interval of length accessible for a given measurement device (a magnifying glass, then a microscope, then a particle accelerator will increase the resolution). These always-new structures prevent one from defining a slope, that is, a tangent to this curve; thus, one is unable to define the speed of a particle which would follow such a trajectory. But a slope can be defined for all given finite values of the resolution: in scale relativity, one thus reintroduces a generalized slope, which is no longer a number, but an explicit function of the resolution.

Such new laws not only change physics at the Planck scale; they also have observable (and thus testable) consequences in the domain of energy currently accessible to large particle accelerators. The problem of the origin of masses and charges of elementary particles can be seen in a new light, as well as that of the value of scales of unification.

Moreover, the consequences of the theory of scale relativity do not only concern the infinitely small: as I will show in closing, *it also allows us to cast a new light on cosmology (the domain of very large scales of length and of the universe taken in its entirety), as well as the general problem of the formation and development of structures*, in particular gravitational structures.

In fact, one of the possible consequences of this approach is that quasi-quantum laws could also be applied to the macroscopic domain. Certainly, it is not a matter of applying quantum theory as we know it now, with all its properties and paradoxes (briefly recalled in the third part). This is out of the question, precisely because our analysis allows us to bring to light everything that, in quantum theory, must be attributed to the particular nature of objects in the microscopic domain (elementarity, identity, nonlocality, indistinguishability, etc.).[110] But this analysis, inversely, allows one also to show that the fundamental equations of quantum mechanics (the Schrödinger, Klein-Gordon, and Dirac equations) have a character of universality.[111] As shown also by the Canadian physicist Garnet Ord, these equations can be obtained without depending on the interpretations of standard quantum mechanics. They can thus have an enlarged domain of application in a different context and with a different interpretation: in such a quantum-like approach (in which macroscopic classical systems may be subjected to a Schrödinger-type regime), if we lose the predictability of individual trajectories, this is largely compensated, as we shall see, by the new capacity of the theory to predict the emergence of structures. In particular, I concluded my 1993 book, *Fractal Space-Time and Microphysics*, by predicting that a fractal medium may play, for the particles which move in this medium, a role similar to that of a fractal space for its geodesics. These particles are then expected, under some conditions of chaos, fractality, and irreversibility, to acquire macroscopic quantum-like properties, but in terms of a macroscopic constant which is no longer the microscopic Planck constant. Much indirect evidence of such a process has been subsequently suggested not only in the astronomical realm, but also in biology[112] and solid state physics[113] and, more recently, direct experimental proofs have been found in laboratory experiments involving turbulent fluids.[114]

Chapter 12
The Concepts of Point and Instant in Physics

Is our current representation of the world complete? As far as classical laws go, the notion of spacetime has been developed with the evolution of ideas in physics. But can we reduce the physical world to the set of positions and instants, and to the motions that tie them together? The classical concept of spacetime can only have physical meaning if the mathematical concept of a "point" is also meaningful.

In the case of time, doubt concerning the existence of an instant, or a temporal point, can be traced back to Greek philosophy and Zeno's paradox. As for space, Kant posed the problem with his antinomies: but Kant went further in showing that the antinomies had bearing upon the infinitely large in addition to the infinitely small. The existence of a symmetry between the statements one can make about these two infinities will become a frequent observation in what follows. One cannot touch zero without touching infinity. Mathematically, they are the inverse of each other, and we will see that this operation of inversion has a very particular meaning in scale relativity.

In its most profound version, the paradox of Zeno's arrow sheds light on the problem. This paradox has two aspects. In the first, one considers that in order to reach a target, an arrow must first travel half the distance, and then half the remaining distance, and so forth *ad infinitum*. In this form, it demonstrates the endless divisibility of space. The mathematical discovery of convergent series, for which an infinite sum of terms can perfectly well be finite, solved the problem. In this way, adding ½, ¼, ⅛, and so on ultimately yields 1 at the limit.

But behind this paradox lies a deeper problem which is still unresolved in the scope of physics today. This problem concerns the concepts of instants and of motion. If the instant, meaning a precise and fixed value of the temporal coordinate, without any incertitude regarding this value, has a physical meaning, Zeno's arrow must be considered at this instant as totally immobile. How does one reconcile the movement of the arrow, known to us by its observed displacement over a finite non-zero interval of time, with this total immobility of the arrow at each individual instant?

For Kant, this line of thinking is still more directly tied to the problem of the infinitely large. Let us first consider the microscopic domain. Is space infinitely divisible? To this question Kant says yes. The appearance of a sudden barrier, an interval of space which one could no longer divide, would have no meaning. But, inversely, does the limit of these successive dissections (the point, the interval of length zero) have meaning? If so, it can only be attained following an infinite number of operations, which cannot actually be brought about. It is similar with the infinitely large. There is no reason why one could not always think of a volume larger than any given volume. But the "limit" of this process (which is in fact *without* limit), an infinite space, cannot practically be realized, even more clearly than in the case of the point zero.

Of course, the statement of these problems has greatly evolved due to the findings of modern physics, in particular with curved spaces which can be finite but without bound in general relativity, as well as the Heisenberg uncertainty principle in quantum mechanics. However, in their essence, the problems remain. For Kant, the fact that each of the two theses is established by demonstrating the impossibility of the other suggests that the question itself is absurd. We will nonetheless see how the theory of scale relativity provides an original answer to the problem of antinomies, in which the two terms are no longer contradictory. The universe can be infinite while possessing a maximal scale, and infinitely divisible even though a lower limit to each scale of length and time can be defined.

Let us continue with the analysis of the concept of position and instant. If the mathematical point can have a proper definition, is it the same for a physical point? Can one bring into existence a true point? The question can also be asked about other geometrical objects such as lines and surfaces, supposedly without thickness. If one wishes to explain what a point is to a child, one can simply draw a dot. But what has one actually done? The sheet, board, or computer screen upon which the point is drawn, even the receptors that are used to detect this image (our eyes in particular) are in actuality characterized by a certain limiting resolution. The dot on the page or screen may be smaller than this resolution, but it still has nothing to do with the mathematical point, which must be strictly zero in extension. One can simply examine it with a magnifying glass to realize that it is in fact an extended spot. Then, one could take a sharper pencil, a pointier instrument, and make a point recognizable as such under the magnifying glass. But using a microscope would also show us its internal structure, and so on until infinity. The mathematical point or line ultimately cannot be physically realized.[115]

Einstein has already insisted on the difference between physics and mathematics:

> It seems to me that when mathematical propositions correspond to reality, they are not certain, and that when they are certain, they do not correspond to reality.[116]

What should we replace the mathematical point, curve, or surface with if they are in actuality inadequate for physical description? The answer is a mathematical tool which includes in its definition itself what the physicist does in practice: one magnifies them more and more with a lens, then an optical microscope, then an electron microscope, then a field-emission microscope, then a particle accelerator. All these instruments change the scale of observation. Toward the larger scales, the use of glasses and telescopes plays a similar role. Yet what experiment and observation has taught us is that never in the course of this type of operation does the strict equivalent of these mathematical objects appear, these objects which we nonetheless use to describe the world. By improving the resolution of an instrument, new internal structures will always appear. But a new geometry now exists, precisely characterized by the existence of structures at every scale: fractal geometry.

Chapter 13
Fractal Geometry

The word "fractal" was coined in 1975 by Benoit Mandelbrot to designate objects, curves, functions, or sets "of which the form is highly irregular and/or fragmented at all scales." Such objects have been studied by mathematicians for over a century. They are characterized by new, often non-integer dimensions, by the appearance of infinities, and by their nondifferentiability (the impossibility of defining a tangent line). Let us explain in more detail what this means.

Topological Dimension and Fractal Dimension

Geometry has always recognized fundamental objects like points, curves, surfaces, or volumes. These objects differ from each other by their dimension. This characteristic can be defined in many different ways, which agree with each other for the simplest geometric objects, but in general differ when applied to the new fractal objects.

The first definition is that of *topological dimension*. This refers to our intuitive concept of dimension: a point has zero dimensions, a straight line one, a surface two, and a volume three. In physics, it corresponds to the number of coordinates necessary to locate a point inside the object.

However, this apparent simplicity was ill-equipped to handle the discovery of strange mathematical sets at the end of the nineteenth century. The Peano curve is one such example (see Figure 5). It is constructed by successive iterations starting from a segment of length one. The first level of construction (the "generator") contains nine segments of length ⅓, and then one reproduces the same structure in each segment, and so forth to infinity.

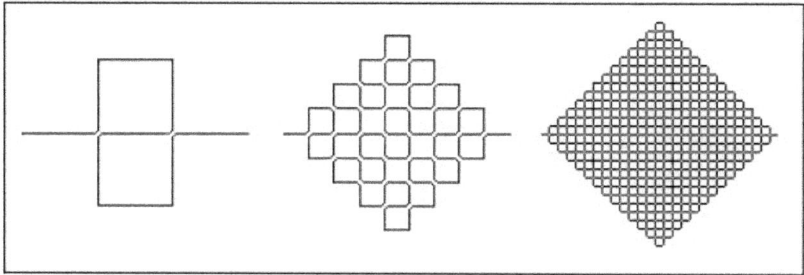

Figure 5 *"Peano curve."* Here we see three levels of construction of this "curve." Its length is infinite. At the limit, it completely fills the surface of a square.

The paradox of such a "curve" is that it allows one to obtain any point on the surface, since at the limit it completely fills the inside of a square! Should one conclude that there is ultimately no essential difference between a curve and a surface? The solution to the problem lies in another property of the Peano curve: while it is true that all points of a surface are touched by this curve, many of them are touched more than once. At the limit of its construction, the curve inevitably cuts in on itself, indeed it must contain an infinite number of multiple points. In other words, if one wants to use this curve as a system of coordinates, the same point must be referred to using many different values of coordinates. This is true of all curves which fill a plane.

This property has allowed one to give a meaningful definition of topological dimensions. Two sets have the same *topological* dimension if and only if one can define a *continuous and one-to-one* transformation between them. Thus, although one can define a continuous transformation between a straight line and the Peano curve, this transformation is not one-to-one, since certain points are reached several times. The "Peano curve," then, is not in fact a curve, but a surface.

The topological dimension of different geometrical objects is thus defined by putting them into correspondence with sets of points (of zero dimensions by definition), of curves (one dimension), or surfaces (two dimensions), etc. This implies that the topological dimension will always be an integer number. As complicated as they may be, it is the same with different fractal objects: one speaks of fractal curves, of fractal surfaces, etc.

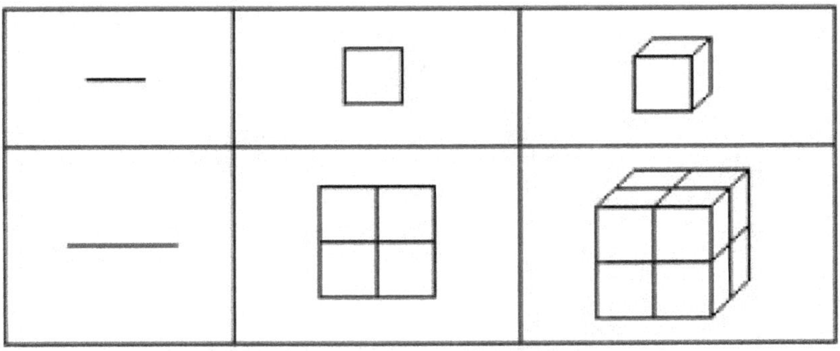

Figure 6 *Similarity of non-fractal objects.* If one doubles the size of a line segment (which is one dimensional), the resulting segment is two times as large. If one doubles the size of a square, which is of two dimensions, one obtains four times the initial square, that is 2^2. In the case of a cube (of three dimensions), the multiplicative factor is 8, or 2^3.

But other definitions of dimension are possible, which in general are called *fractal dimensions*. Among them, one plays a central role in the study of fractals, the *similarity dimension*. It is based on the following observation: when one applies a homothetic factor n to a line segment (that is, when one enlarges it by n times), one obtains a new line segment n times as large; applied to a square, one obtains a new square with a surface n^2 as large; for a cube, the volume is multiplied by n^3; for a "hypercube" (generalization of the cube beyond three dimensions) having a topological dimension D_T, the multiplicative factor is n^{DT}, etc. (see figure 6). The exponent used in the factor of enlargement is the similarity dimension. In all cases of standard geometry, the similarity dimension coincides with the topological dimension.

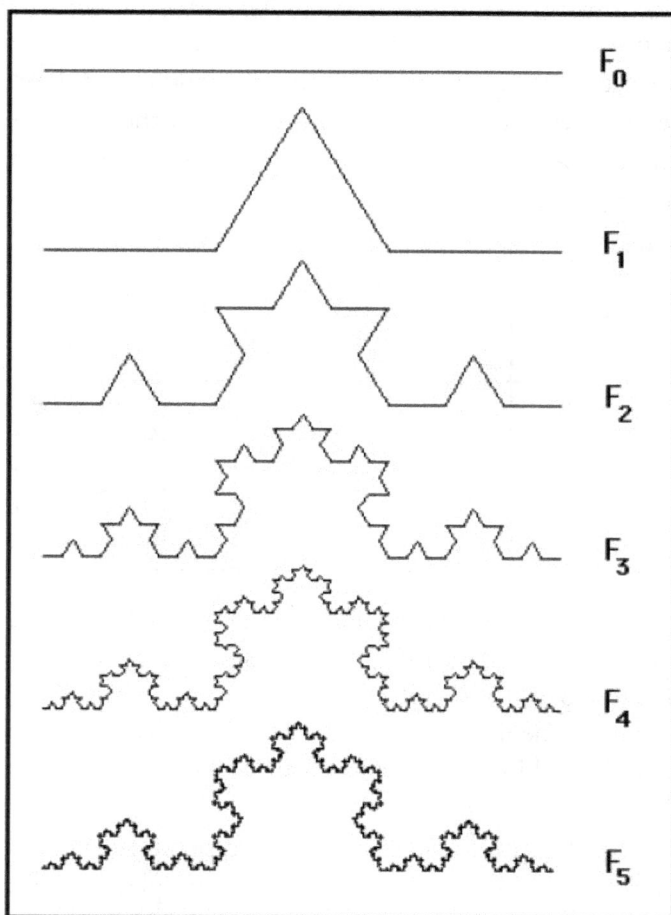

Figure 7 *"Koch curve."* This curve is constructed starting from a "generator" F_1, which contains four segments of length ⅓, by replacing F_1 by F_1 itself, but at smaller scale, and then continuing this process to infinity. Its length is infinite (each iteration multiplies the length by 4/3). Its fractal dimension is 1.26. This curve is not differentiable, since the tangent of a point will differ according to the level of construction and is not, in general, defined at the limit.

However, if one enlarges a self-similar fractal by a factor q, one can obtain p versions of the initial fractal, without p being a power of q. This means that the similarity dimension can now be non-integer. For example, the famous "Koch curve" (see figure 7), magnified three times, contains four versions of the initial curve! The similarity dimension, which is one of the examples of the new *fractal dimensions*, is then given by the ratio between the logarithms of p and q.[117] That of the "Koch curve," which is equal to log 4 / log 3, about 1.262, is greater than its topological dimension (equal to one).

Everything behaves as if a fractal curve had a kind of thickness which makes it something in between a line and a surface. One can understand this by comparing a straight line, which is an infinitely narrow curve with one dimension, with the Koch curve above (of fractal dimension D = 1.26), and then with the zigzag fractal in figure 8 (fractal dimension 1.5), and finally to the Peano fractal which fills the plane and has a fractal dimension of 2.

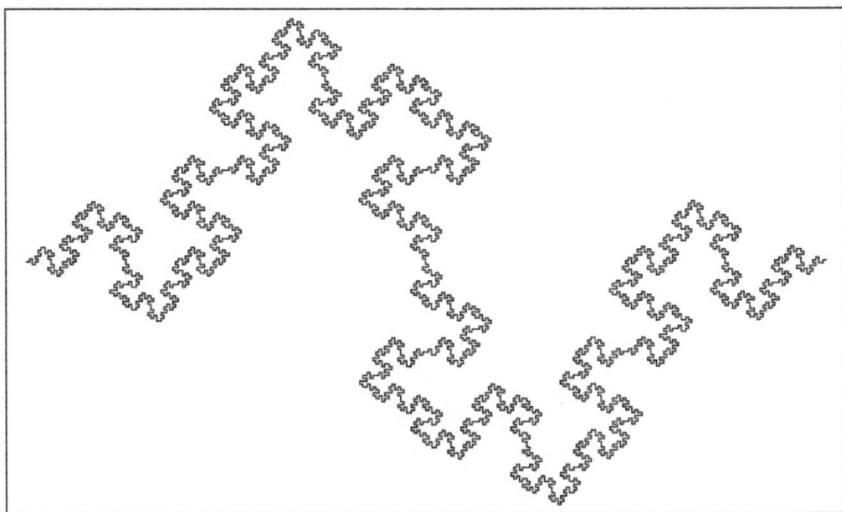

Figure 8 *Fractal curve of dimension 1.5.* Its generator is composed of 8 segments of length ¼.

An important property of fractal curves is their "nondifferentiability." This can be seen, for a curve in a plane, by the impossibility of defining a tangent when one progresses to the limit of the iterative construction of the fractal, as in the Koch curve above. In the case of a *fractal function*, that is, where each value on the x-axis corresponds to a single value on the y-axis, nondifferentiability exists as a divergence of the slope (meaning that the tangent to the curve is always vertical): in this case, the derivation does not exist in the usual sense of the term, not only because it is undefined due to fluctuating endlessly under changes of scale, but also because it is infinite at the limit (see figure 9).

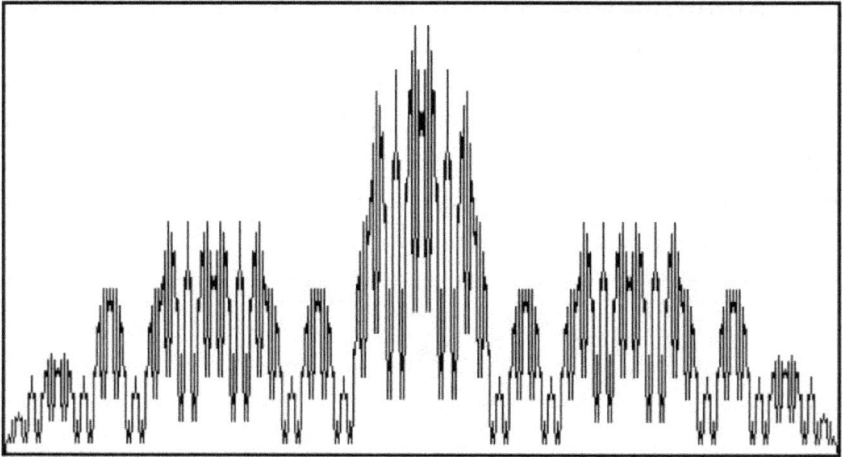

Figure 9 *Fractal function*. This function is obtained by projecting the fractal curve of the preceding figure on the horizontal axis. The derivative of this function tends to infinity at each of its points.

Another type of fractal dimension is the *covering dimension*. To understand the nature of a fractal curve, it is possible to consider it at all the different possible resolutions. This can be done by covering it with "balls" of variable radius ε. To simplify the example, we will consider the "resolution" to be the radius of this ball, which in covering the curve creates a smoothing effect.[118] One can also see this radius as a kind of uncertainty or an error bar on the positions of points on the curve.

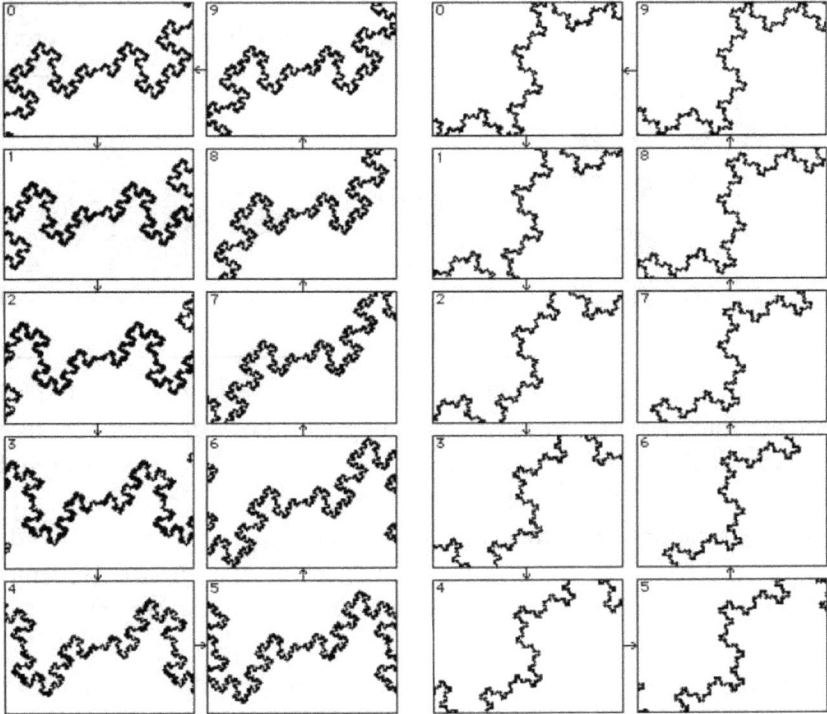

Figure 10 *Zooming in on two fractal curves.* Each of these two series of images can be made into frames of a film. Since the fractals are self-similar, the series loop on themselves, image ten being identical to image 0.

What characterizes a fractal curve is the appearance of new structures when the resolution improves, that is, when the covering ball becomes smaller and smaller. These supplemental structures can lead to a growth without limit in the length of the curve when the interval of resolution tends to zero. In the simplest case of self-similarity, this growth follows a power law, the exponent of this law characterizing the fractal covering dimension.[119] If one continuously zooms in on such a fractal curve, one will periodically come across the same structures (see figure 10).

In the case of a fractal surface (see figure 11), the measured area tends to infinity over the course of successive zoom-ins, due to the continual appearance of new fluctuations upon this surface. The same goes for higher topological dimensions (volumes, hypervolumes, etc.).

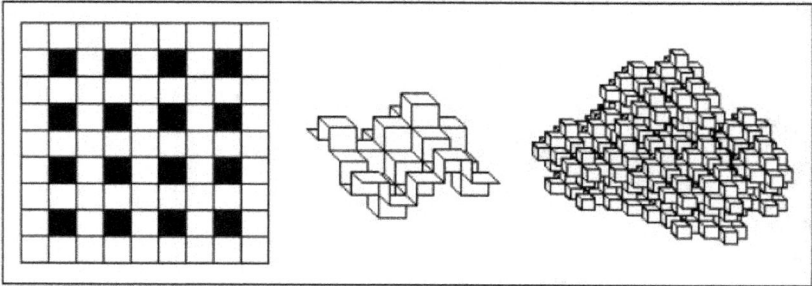

Figure 11 *Fractal surface.* One can construct a continuous fractal starting from the folding of a generator containing gaps (at left). One obtains the generator in the middle, composed of 65 squares with sides of length 1/5 (fractal dimension 2.59). The figure on the right gives an idea of the next level in the construction, in which each square of the generator is replaced by a copy of itself at smaller scale. Only part of the surface is depicted (corresponding to the five cubes in the upper right of the generator).

Nevertheless, we must note that most of the fractal objects encountered in nature do not truly display structures at *all* scales. They are geometrically characterized by the existence of lower and/or upper scales of transition beyond which they become standard, unstructured objects. As we will see in the following, it will be the same for fractal spacetimes which can be constructed to account for quantum behavior. In this latter case, the transition between fractal behavior (at relatively small scales) and non-fractal (at larger scales) marks the transition between quantum and classical behavior.[120] This type of transition, which does not take place in spacetime, but in the new space of resolutions (space which must be introduced to allow a complete description of a fractal object), can be identified as a spontaneous "symmetry breaking" of the laws of scale. Strictly speaking, the quantum spacetime is fractal at all scales, but the combination of the laws of scales with the laws of motion (i.e., of the two relativities, of scale and of motion) leads to the appearance of an effective transition from the quantum realm (dominated by scale laws) to the classical realm (dominated by motion laws).

Two examples will illustrate this property. One of the best known instances of a fractal curve is the coast of Britain, whose length grows when one measures it using maps of larger and larger scale. This growth ends when one reaches the scale of the shore itself.[121]

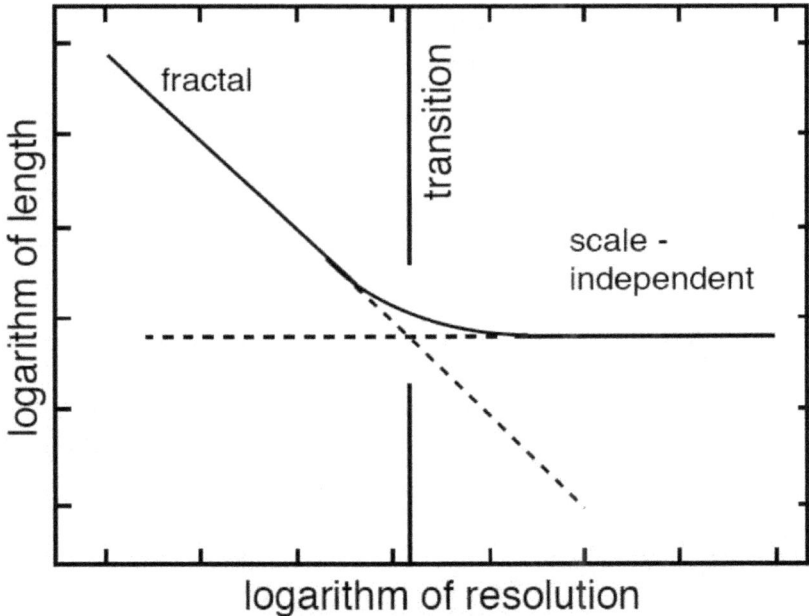

Figure 12 *Transition between fractal behavior and scale-independent behavior.* The length of a curve of fractal dimension 2 is plotted as a function of the resolution of measurement. The units are logarithmic: the same distance on this diagram corresponds to a constant ratio (the values 1, 2, and 3 correspond to an increase of 10, 100, and 1,000 times the initial length). At large scale, the surface upon which this curve is drawn is relatively smooth, so much that the length of the curve does not depend on its resolution; however, at smaller scales, aberrations in the surface appear more and more and the length of the curve rapidly grows.

Inversely, let us consider the surface of the sheet of paper upon which these lines are printed. If we measure its area with a resolution of a centimeter, then of a millimeter, and even of a tenth of a millimeter, we will obtain the same value (with greater and greater precision): at these (relatively) large scales, there is no variation of the area as a function of the resolution. But what about smaller scales? As the magnification increases, the irregularities of the paper will appear more and more; the total area must take these structures into account, and then the structures within the structures. The area will thus increase with the diminution of the scale of resolution. Here we have a fractal object at small scale which is not fractal at larger scales, with a rather quick transition between the two behaviors (see figure 12).

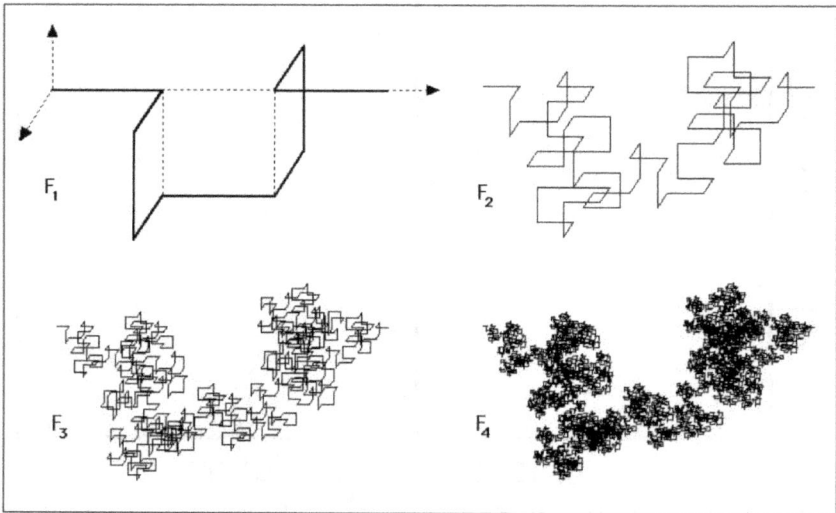

Figure 13 *Fractal curve in space.* The generator of this curve, which exists in a three-dimensional space, is composed of 9 segments of length ⅓, so that its fractal dimension is $D = 2$. Four successive levels of construction by iteration are shown. Such a curve possesses an infinite length and zero volume, but has a finite "area," even though it is not a surface!

In the simplest cases, the different definitions of fractal dimensions coincide. This result allowed Mandelbrot to introduce the general concept of "fractal dimension," which corresponds in many real situations to a covering dimension or a similarity dimension. Whatever it may be, I will here use the word fractal, in particular when it will be applied to spacetime, in a very general sense: that of an explicit dependence of metric quantities (lengths, surfaces) as a function of spatiotemporal resolutions, which may diverge at the smallest scales.

Let us note here that it would be incorrect to define fractals as sets characterized by non-integer dimensions. For example, one can perfectly well construct *curves* (therefore of topological dimension 1), which do not cross themselves, having a fractal dimension 2, within a space with three dimensions (see figure 13): as we shall see, these are just the characteristics of the geodesics of a fractal spacetime that lead to standard quantum mechanics.

Curves possessing this type of property play an essential role in the new approach to quantum phenomena which we will discuss in the rest of this book. The measurement of these curves is an area, even though they do not at all resemble surfaces. Inversely, if the fractal dimension of a continuous fractal is equal to the topological dimension of the space which contains it, one knows that it will cross itself at an infinite number of points (we say it possesses multiple points).

It is also possible to imagine generalizations of ordinary fractals, in which the fractal dimension becomes variable as a function of position (see figure 14) or of scale. This latter class of curves plays an important role in physics, as we shall see below.

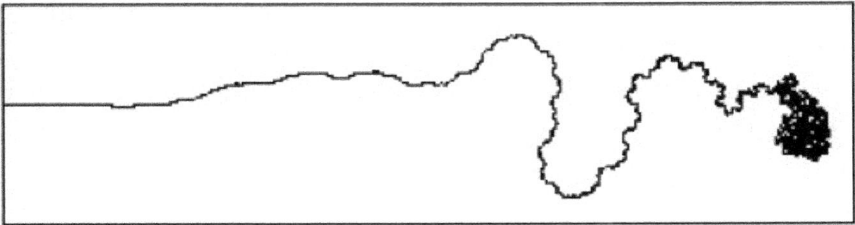

Figure 14 *Fractal curve of variable dimension.* The dimension is at first equal to one (left part of the curve), then increases regularly to reach 2, the value for which it fills the plane.

Where Do Fractals Come From?

The extraordinarily frequent appearance of fractal structures in natural systems (physical and biological) is now an undeniable observational fact. Mandelbrot gave a large number of examples (coast of Britain, mountainous surfaces, crystalline structures, distribution of galaxies, Brownian motion, turbulence, lung structure, etc.), and further research has added yet more examples (asteroids, lunar craters, solar spectrum, brain, circulation, and digestive systems, hadron jets, growth phenomena). The list is now so long that it is practically impossible to be exhaustive.

The now-definitive confirmation of the universality of fractals demands, in a more urgent manner than in the past, an explanation of their physical origin. From what underlying laws of nature do they emerge? One might answer this question from a number of possible angles. But the ultimate response might be that *fractal geometries are simply more general than Euclidean and curved geometries.* Fractals would then be, in the case where they occur at a fundamental level (involving spacetime itself), the structural manifestation of the primary nondifferentiability of nature.

Thus, to recognize that fractal phenomena dominate non-fractal ones leads us to recognize finally that the laws of nature are not governed by differentiability, a fundamental axiom which underlies all of classical mechanics. However, before coming to this general argument, some examples can be given of the physical processes naturally leading to the emergence of fractal behavior.

Optimization Under Constraint

One can obtain fractal structures in terms of a *process of optimization under constraint*, or more generally of the optimization of several quantities, sometimes apparently contradictory. Suppose, for example, that the evolution of a system leads to a maximization of surface area (which is the case in the process of exchange, as in the lung) while minimizing the volume. A solution which optimizes the two constraints is a fractal of dimension greater than two, but smaller than three (it would correspond to a surface tending toward infinity and an infinitesimal volume).

One of the simplest cases of fractal objects (which should not be considered the only case) is that of self-similar fractal objects. These are sets for which one observes the *same* structure after successive magnifications. However, for the mathematical objects which have been most studied thus far, one does not find the *exact* initial configuration unless considered at discrete values of the dilation factor. In other words, when one analyzes them in terms of the variable of resolution (as a logarithm), a self-similar fractal is a *periodic* system in the new dimension of scale.

This "discrete scale invariance" has recently been developed and applied in numerous domains, in particular by Didier Sornette and his collaborators, to earthquakes or stock market crashes, but also by Chaline, Grou, and myself to the evolution of species and of human societies,[122] and to embryogenesis.[123] What might be the origin of such a discrete scale invariance? Under the hypothesis where fractalization does indeed arise from a law of optimization, one can interpret it in the following manner: after a first generic dilation which allows for optimization, there comes up again a new blockage in the system, which is thus *returned to the preceding state*, except for a scaling factor. The problem being the same, the solution is also similar, and therefore a new iteration can operate (see Fig. 15).

Let us illustrate these two aspects of fractals, that is, multiple optimization and the return to a preceding state, by a simple example. Suppose that a system needed to increase the energetic intake coming from the environment, which is done with a membrane. This is proportional to the surface of the membrane; thus the system necessitates the maximization of a quantity of topological dimension $D_T = 2$. The simplest solution would be to exchange energy across the external surface limiting the body. Optimization would thus lead to an increase in size. However, the end result of such a process is negative: increasing the size by a factor ρ increases the surface by ρ^2, while the volume increases by ρ^3, so that the energy per unit of volume *decreases* by $1/\rho$. Moreover, the thermal energy (of heat) losses take place across the external surface, so well that it is not possible to increase it by too much (this, with the constraint of gravity, leads to a well-known limit to the size of living organisms on Earth).

Thus the problem is now posed in a different manner: is it possible to increase the surface of exchange without increasing the volume of the body (or equivalently, its external surface)? The only solution is clearly to enlarge the *interior* surface of the body by "invagination" (see Figure 15).

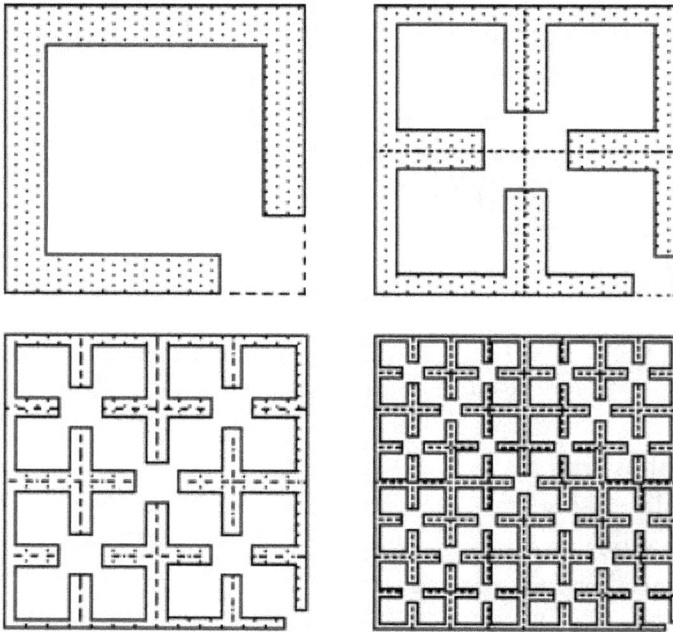

Figure 15 *Model of self-similar growth.* The fractal nature of an exchange *surface* (here, viewed in cross section and reduced to a line) allowing the growth of its area without increasing the exterior volume. It is thus with the lung, whose total exchange surface is more than 100 m².

Still, this increase will soon encounter another limit, when the internal growth is limited by the external surface. But when this limit is attained, the problem posed is brought back to the preceding problem.

The solution is the same: new invaginations, at smaller scale. Self-similarity is here assured by the principle of causality. The same causes produce the same effects, and one expects that fractalization will continue to smaller and smaller scales with self-similar structures. The process stops when other constraints are applied to the system (in our example, the thickness of the membrane and diameter of various openings). The lung, with its twenty levels of fractalization (in base two), from the trachea to the alveoli, is an example of a natural system where such a model could describe it as a first approximation.

Renormalization Group

Another path toward the emergence of fractal properties rests on the form of the renormalization group. It is an approach for multi-scale phenomena, developed mostly by Kenneth Wilson. These methods had initially been developed in the domain of quantum field theory.

The approach of the renormalization group consists of locally describing the system at *small* scale, and then dilating the system by a certain scale factor. One then studies the way in which various quantities, fields, coupling constants, etc., have changed (have been "renormalized" in this transformation). One is thus brought back to the preceding problem and the process can be iterated. The advantage of this method is that the number of steps allowing one to move from elementary structures to global description is now given by the logarithm of the number of elements, and not by the number itself.

Here is an example of a problem which this type of method aims to resolve. The global magnetization of a material results from the orientation of elementary spins of all its atoms. How can one calculate this considering that the number of atoms is of the order of 10^{25}? It seems impossible to manage such immense numbers. With the method of the renormalization group, one considers all the possible groupings of a small set of elementary quantum spins, then one iterates with groupings of groupings, etc. If the spins have been grouped in blocks of ten, twenty-five steps suffice instead of 10^{25}!

The similitude between fractals and the renormalization group is established by itself: iteration and scale dependence are present in the two cases.[124] One must nevertheless mention an important difference. In going from one scale to the next one (larger), one usually replaces the information about the system with an average, so that it is not possible to return to the smaller scale. In other words, there is no inverse transformation in the renormalization group: it is, mathematically speaking, a semigroup. On the other hand, fractals are often constructed starting from a large scale and moving toward smaller scales. One defines a generator (which is the elementary structure which will be reproduced by iterations), then one constructs smaller and smaller structures by the successive application of the generator after it has been reduced in scale. In this sense, fractals can be conceptualized as a sort of *inverse transformation* of the renormalization group.

Dynamical Chaos

An important area of research in recent years in which fractals play a leading role is that of dynamical chaos. What is this chaos? It is random behavior which is produced in certain deterministic systems. It thus has nothing to do with the usual meaning of the word chaos, which refers to confusion, complete disorder, and absence of law. Chaos is a property of (seemingly) deterministic systems, those which are adequately described by laws, in general those of classical mechanics. Our fascination with it perhaps comes from this paradox, discovered by Henri Poincaré: "A system described by perfectly deterministic laws can have a behavior arising from chance!"

How is this possible? The answer has to do with the nature of laws in modern physics. These laws describe relations, constraints between physical quantities, which are expressed mathematically by differential equations. However, knowing an equation does not automatically mean understanding its solutions.

Chaos is characterized by a high sensitivity to initial conditions. Two initially close trajectories, both solutions to the same equation, can diverge from one another extremely rapidly. When this divergence is an exponential function of time, the final separation can quickly reach enormous values, even though the initial separation was tiny. The determinism of the equation then becomes illusory. It corresponds to an idealized situation where one can know the position and the speed of an object at any given instant with an infinite precision. In reality, the existence of an inevitable uncertainty concerning these quantities implies, for a chaotic system, the existence of a temporal horizon of predictability, beyond which the precise evolution of a particular trajectory becomes impossible to predict.

Even in this case, a partial theoretic prediction can still be possible. This is what allows one to speak of "order in chaos." To understand this, let us first consider a non-chaotic system, one that is completely predictable, such as a swinging pendulum. A pendulum will oscillate less and less until it lies at rest at its lowest point. This final condition (the lowest point and zero velocity), independent of the initial conditions (how one has moved the pendulum and the speed at which one pushed it), is called the *attractor* of the system. This evocative term should not make one believe in some kind of finalism: it is a case of a deterministic system, governed by causality.

Yet many chaotic systems can also be characterized by attractors, which have been called "strange" by Ruelle and Takens due to their great complexity. Such chaotic attractors are often fractal (see figure 16).

Celestial mechanics offers numerous other examples of the relation between chaos and fractals: fractals are not explained by chaos, or chaos by fractals, but the same underlying phenomenon (the resonances between orbital periods) gives rise to both the fractal structure and to chaotic trajectories, which ultimately leads to unpredictability.

Consider the three-body problem corresponding to two massive bodies and a test-particle. It is thus, for example, with the combination formed by the Sun, Jupiter, and an asteroid. In such a case, the source of chaos and its fractal properties begin to be well understood. Chaos arises from the resonances between the orbits, which are produced when the periods of Jupiter and the test-object are fractions composed of small integer numbers. Take the example of a body having a period double that of Jupiter (resonance 2 : 1). Every two revolutions, the body passes extremely close to Jupiter and is submitted to a violent "kick" which perturbs its trajectory. At the limit, it can finally be ejected, if its trajectory, from the point of its greatest eccentricity, is led to cross the orbits of other planets: such a phenomenon allowed J. Wisdom to explain several of the so-called Kirkwood gaps in the asteroid belt found between Mars and Jupiter.

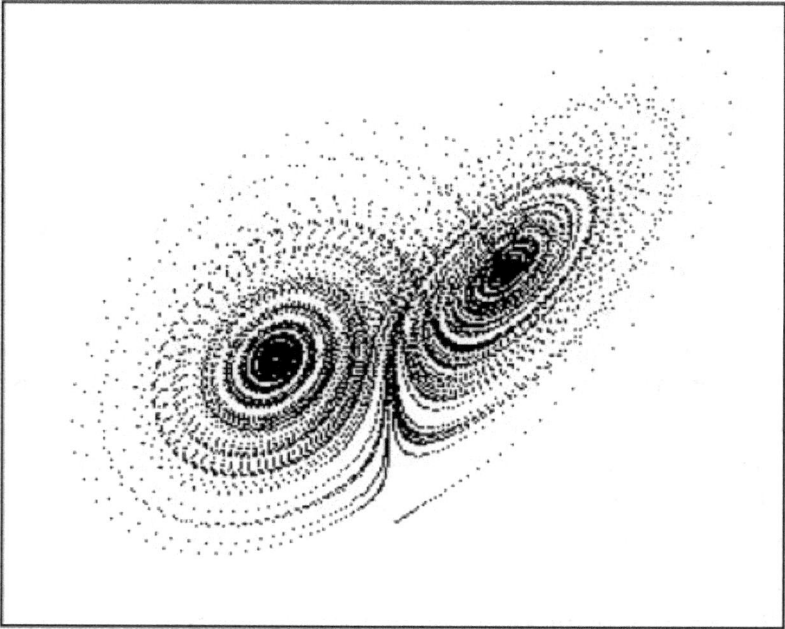

Figure 16 *Lorenz's chaotic attractor.* This attractor corresponds to the evolution of certain variables of a simplified model of atmospheric convection (the rise of hot air). If one starts from whatever conditions, the system will always evolve toward this attractor composed of two layers. Once on the attractor, the state of the system will advance by turning around one of the centers and occasionally jumping from one layer to the other in an unpredictable manner.

Over very long scales of time, it is, at the limit, the set of the distribution of rational numbers, which forms a fractal dust, which could intervene to determine the distribution of asteroids between Mars and Jupiter (see figure 17).

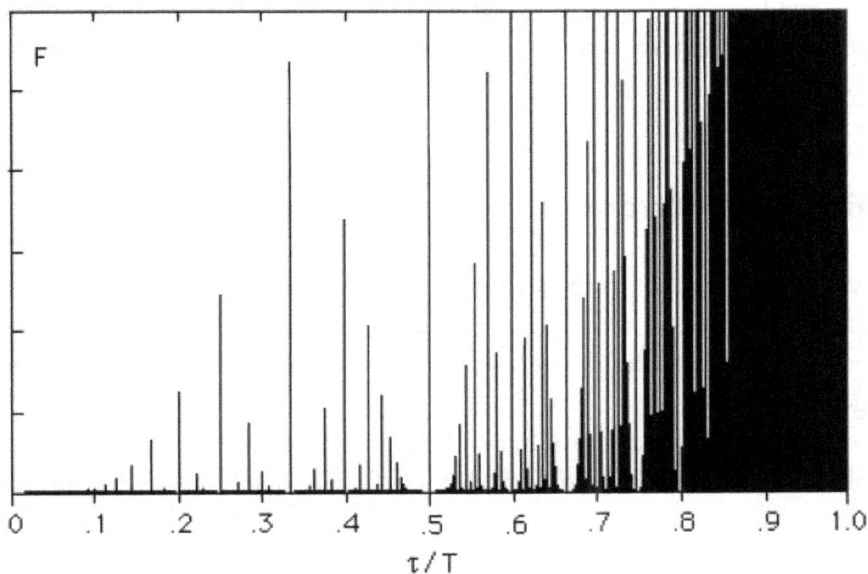

Figure 17 *Fluctuations of Jupiter's force.* Fluctuations of the average force exerted by Jupiter on another body of the solar system. I have plotted the amplitude of this fluctuation as a function of the ratio of the period between the body and Jupiter (which varies as the distance of the Sun at the power of 3/2). Only the simplified case of circular orbits is considered. The fluctuation becomes significant for rational ratios of periods. It is thus a fractal function corresponding to the distribution of rational numbers among the reals.

But chaos theory has led to an even more dramatic result: it has been shown by Jacques Laskar that the solar system is chaotic and that the positions of planets could no longer be predicted over 100 million years. The chaos is so strong for a planet like Mercury, that, according to this description in terms of classical equations of dynamics, it could be ejected from the solar system within 3 billion years.

However, as we shall see in what follows, the new scale relativity approach leads to a different conclusion: its application is founded on the loss of predictability on very long timescales due to chaos, which may allow a transformation of the classical equations into a new, quantum-type form. Under this "Schrödinger regime," one finds theoretically that planetary systems can be stabilized provided the planets follow quantized orbits. This expectation has just been verified in our solar system and in some recently discovered extrasolar planetary systems.

Beyond Differential Geometry

When it is applied to the structure of spacetime, the concept of fractals can be justified in a much more fundamental manner: it can be defined as a nondifferentiable generalization of Riemannian (i.e. curved) geometry.

A fundamental theorem connects fractals to nondifferentiability. It allows us to introduce the notion of fractal space, more generally of fractal spacetime, not as a supplemental hypothesis in relation to classical space, but on the contrary as the consequence of *abandoning* a hypothesis, that of differentiability. This theorem states that *a continuous curve that is nowhere (or almost nowhere) differentiable is fractal.* The word "fractal" is here used in the sense in which the length of such a curve is explicitly dependent on the resolution with which one considers it and tends toward infinity as the resolution tends toward zero.[125] One can easily generalize the proof to a surface, a volume, and more generally a continuous space or spacetime.

As we shall see, the concept of fractal space (more generally, fractal spacetime) will play an essential role in the future developments of fundamental physics. Up to now, even if the words "fractal geometry" have been abundantly used, the mathematical domain of fractals has not yet attained a status comparable to that of Euclidean or Riemannian geometry. The concept of fractals in its current usage characterizes certain natural phenomena or objects immersed in an underlying Euclidean space.

Applying the concept of fractals to spaces is a different story. To understand this, let us return to Gauss's discovery of curved geometries. This discovery became effective when Gauss was able to characterize a surface using completely *intrinsic* methods, in particular with the definition of invariants which did not depend on coordinates traced upon this surface (as with latitude and longitude on the terrestrial globe). This allows one to describe already-known surfaces, such as spheres, from inside the two-dimensional surface itself (without referring to the three-dimensional volume of the sphere under the surface). But this intrinsic definition was important mainly because it allowed one to prove the logical existence of surfaces which cannot be immersed in a Euclidean space, such as hyperbolic spaces. After opening this "Pandora's box," Gauss partially shut it again by limiting the new geometry to *locally Euclidean* spaces. This hypothesis was conserved by Riemann in his more general construction of spaces possessing more than two dimensions. Gauss's hypothesis also played a fundamental role when Einstein used Riemannian geometries as a mathematical tool in his construction of the theory of general relativity. In effect, to suppose that spacetime is locally flat implies that around a point, the principles of special relativity apply, within which the laws of physics are those of free inertial motion. Thus, Gauss's hypothesis is the basis of the *mathematical transcription of the equivalence principle*, according to which gravitation disappears in a freely moving frame of reference. In other words, since gravitation is the manifestation of curvature, the local absence of curvature means the local absence of gravitation.

Today, the Pandora's box has been opened anew with fractal geometries. One can consider the fractal hypothesis as the abandonment of Gauss's hypothesis. In the case of fractals, instead of finding flatness at small scales, one sees the constant appearance of new details as the scale becomes smaller and smaller. This implies that the curvature of a fractal space tends toward infinity as the resolution δx tends toward zero. One can construct fractal surfaces which are flat at large scale, and whose curvature at the infinitesimal level is everywhere infinite, these infinities, alternately positive and negative, being themselves distributed in a fractal manner.

Thus, what I will call fractal spacetime throughout this book is, by definition, a geometry which is both nondifferentiable and explicitly dependent on scale. Such a concept goes well beyond ordinary spacetime and cannot be reduced to the latter, due to the appearance of structures at every scale.

In such a framework, the question "Why should spacetime be fractal?" gains a similar status to that of the traditional question "Why is spacetime curved?" In both cases, the answer that one can suggest is that curvature, then fractalization are simply the expression of abandoning an axiom which had heretofore been implicitly made in a simplifying manner, that is, a process of moving toward a greater generality. In these conditions, the new theory cannot be in contradiction with the previous one, since it necessarily contains the latter within itself, being an extension of it. In the same way that curved geometries always contain as special cases Euclidean geometries, or that Einstein's equations can allow flat spaces as solutions, continuous "fractal" spaces (in this extended meaning) include differentiable and nondifferentiable geometries and therefore contain curved spaces as special cases. It is a matter of taking into consideration all continuous spaces, whether they are differentiable or not differentiable (and therefore fractal).

We are certainly very far, within the new fractal approach, from the successes obtained by general relativity. We can all the same set out the goal of such a theoretical effort: this would be to find a system of equations which would be *covariant under the continuous (but not necessarily differentiable) transformations of coordinates.* The theorem stated above, which connects nondifferentiability and scale dependence, implies the necessity in this endeavor of taking into consideration not only changes of motion, but changes in scale as well.

One can hope that such a program will provide, as an inevitable consequence, the description of fractal structures of spacetime. But one must keep in mind that such structures, imposed by physics at a fundamental level, would in all likelihood possess different properties from the fractals which we know of today. We will see in particular that quantum behavior itself would be, in this approach, a manifestation of this underlying fractal geometry.

From Fractal Objects to Fractal Spaces

Fractal curves, with their infinite length, their fractal dimension greater than one, their thickness, and their absence of slope (that is, their nondifferentiability), constitute geometric objects which, within a plane, occupy an intermediate position between lines and surfaces. In space, they can become intermediaries between lines and volumes. One can thus construct curves of fractal dimension 2, whose length is infinite, the volume infinitesimal, but whose "area" is finite. The measure of the "contents" of such an object can be expressed in square meters (like surfaces), even though it is an authentic curve.

In general relativity (and in the Riemannian spaces which constitute the geometric tool of Einstein's theory), the notion of coordinate system is generalized to curvilinear coordinates (Einstein used the image of a "mollusk" of coordinates). The "axes" of such a system are no longer the rectilinear angles of a Cartesian frame of reference, but can become curves (traced in spacetime, which includes accelerations). Nevertheless, these curvilinear reference systems, as well as the transformations which connect them, remain differentiable.

With the concept of the fractal curve, one can imagine going further and placing oneself within a system of fractal coordinates. Would this be enough for a new physics to appear? The answer is no. Here again, the example of general relativity is illuminating. In placing oneself in a curvilinear system (accelerated), the forces of inertia appear which resemble gravitation (this is the equivalence principle). Thus, the laws inside a rocket accelerated at 1 g (the acceleration of Earth's gravity) are, locally, at all points similar to those of the fall of bodies on Earth. But true gravitation—that is to say, a manifestation of the geometric curvature of spacetime—is different in essence. It corresponds to something which, other than locally, cannot be cancelled out by a choice of coordinate system. From this point of view, a curve is not "curved": as complex as it may be, it can be unbent into a straight line without being broken. On the contrary, a curved surface (for example an orange peel) cannot be applied to a plane without being torn or folded (which makes it impossible to render a map of the world without deformation or omission). It is this property that we call curvature and which is manifest in the form of gravitation.

It is the same for fractal curves which, from the point of view that interests us here (the appearance of a new field, a manifestation of fractal geometry), no longer have curvature, nor any irreducible geometric property. In effect, a fractal line can, like a differentiable curve, be unbent until it becomes a straight line. The difference is that a portion of a differentiable curve is of finite length and becomes, straightened out, a finite segment, while a fractal curve, having an infinite length, becomes an infinite line once unbent (see the fractal curves shown in figures 4-14).

Nevertheless, if one were to try to flatten a fractal surface to a plane, one would tear it at each of its points! One can be convinced of this starting from the construction, depicted in figure 11, of a continuous fractal surface made from the folding of a flat fractal with holes. Its curvature is effectively everywhere infinite, with a fractal distribution of positive and negative signs.

It is this type of irreducible property in the choice of coordinate system which could allow for the appearance of a universal "field." I will suggest below that one of the manifestations of such a field is quantum mechanics itself.

Chapter 14
The "Missing Link" in Quantum Theory

Comparing the structure of current quantum theory and that of general relativity reveals the profound incompleteness of the former, as we will see. It is not a matter of incompleteness in the sense where a return to determinism would be necessary, but a profound conceptual gap. What would this missing concept be in quantum theory? The answer, proposed by the theory of scale relativity is: *spacetime*.

The fundamental equations of general relativity correspond, in fact, to the geometric description of a dynamic spacetime, tied to its material and energetic contents. The conceptions of Leibniz, then the analysis of Mach reinforced by Einstein, have led to the physical impossibility of an absolute spacetime independent of its contents.

But what is the spacetime of quantum theory? It is, according to the current quantum theory, the flat spacetime of Galilean relativity or, at high energies, that of special relativity. Both are absolute by essence. However, at very small scales, all objects encountered and all the physical properties of these objects have a quantum nature. How is it that the spacetime of which they are the contents is not affected? How can it remain classical, absolute and independent of the quantum properties of the matter that it contains. Here is a flagrant contradiction of the current description of the microscopic domain with respect to the historic evolution of ideas in physics.

This point of view is confirmed by the nature of the fundamental equations of the two theories. Those of general relativity describe the geometry of spacetime; the equations of the motion of bodies are deduced from the equations of geodesics (that is, the shortest possible lines according to proper time). Nevertheless, those of quantum mechanics are equivalent to equations of trajectories (more precisely, to a set of paths, which cannot be identified with the classical concept of trajectory of a particle). Could there not exist an underlying spacetime, of which the fundamental equations (Schrödinger's equation, Dirac's equation) would describe the geodesics? Such is the fundamental question of the new scale relativity approach.

The Nature of Quantum Spacetime

The problem comes down to asking oneself what kind of geometry could give rise to quantum effects, in a similar manner to how curvature manifests itself as what we call gravitation. For such an endeavor, lessons from the history of science are an essential aide. Since Einstein established his theory of gravitation (1915-1917), numerous attempts at extending the theory have been undertaken. One of the principal goals of these attempts was to construct a unified theory of electromagnetism and gravitation. But Einstein's deep intention over the course of his multiple attempts, which he pursued up until the end of his life, was also to take into consideration quantum effects. It is thus that he wrote in a letter to Louis de Broglie at the end of his life (February 15, 1954):

> In truth, like you I am convinced that we must find a substructure, a necessity that current quantum theory cleverly hides with the application of the statistical form.
>
> But for a long time I have been certain that we will not be able to find this substructure by a constructive method beginning from (empirical) behavior of known things, since the necessary conceptual leap exceeds human

ability. It is not only because of the futility of numerous years of effort that I have arrived at this opinion, but also by my experience in the theory of gravitation. The equations of gravitation could have been discovered solely on the basis of a purely formal principle (general covariance), that is, on the basis of the conviction that the laws of nature have the greatest logical simplicity imaginable. As it was obvious that the theory of gravitation constituted only a first step toward the discovery of the simplest possible general field laws, it seemed to me that first of all this logical method must be pursued to the end before being able to hope to arrive at a solution to the quantum problem as well.[126]

The failure of these attempts allows one to convince oneself of the impossibility of obtaining quantum effects starting from the geometry of differentiable manifolds. The extensions of general relativity considered by Einstein (and many other physicists, such as Schrödinger, Kaluza and Klein, Jordan, Weyl, etc.), even if they introduced new elements such as supplemental dimensions or torsion in addition to curvature, all remained within the framework of differentiable spacetimes.

But the conclusion (that many physicists drew) according to which all geometric attempts were doomed to failure was premature. It was making the implicit hypothesis that there was nothing beyond Riemannian geometry or its known extensions. The only valid conclusion is that an eventual deeper vision of the nature of spacetime in microphysics will not be obtained unless new concepts are introduced. Moreover, an essential fact can serve to guide us in the search for new geometric tools: the spatiotemporal approach, being a description of the framework and not only the objects within a pre-established framework, cannot be based on particular fields, but instead on the properties of matter and radiation which have a character of universality.

Toward a Nondifferentiable Spacetime

A new area of research has been opened. Why must spacetime be differentiable? No proof exists, either theoretical or experimental, of its differentiability. Certainly, all of celestial mechanics and its tremendous successes demonstrate, if there was any need, that the hypothesis of differentiability has no reason to be abandoned within a large physical domain. But this success does not extend to scales of length and time which are much smaller or greater than those where experimental and observational results have been able to be obtained.

Yet the mathematical statement of the existence of a derivative for a variable (in particular those of position and instant) is precisely an asymptotic statement of a very small scale. For example, the existence of a derivative over the course of time, that is to say, a velocity, supposes the existence of a limit for the ratio between the variation of position and the time interval when the latter tends toward zero. However, it is precisely by investigating the microscopic domain that quantum effects have been demonstrated. Classical mechanics, differentiable by nature, no longer applies at small scales, where another form of mechanics, quantum, applies. Is this not exactly because the hypothesis of differentiability finally shows its limits?

One might object to this point of view that the impossibility of defining a velocity should also show effects on classical scales. But the physically measured velocity is very different from the mathematical derivative. In the mathematical definition of velocity, the ratio of a length interval and a time interval which are zero at the limit must be taken into account. On the other hand, the physical measurement of velocity can only be performed at finite non-zero intervals of length and time. Measuring a speed in classical mechanics necessarily considers only intervals of length and time which stay within the classical domain.

Nothing, then, prevents the nondifferentiability of space at very small scales, since nondifferentiability could well be unapparent at the macroscopic domain. It is possible to go further: not only is it not excluded, but numerous arguments seem to argue for its existence.

Einstein himself glimpsed the possibility that abandoning differentiability was a key for understanding the quantum domain. In a 1948 letter to Pauli (of which an extract has already been cited above), he wrote:

> If the Ψ function does not completely describe the real situation of an individual system, there should still be a complete description, and we must find it. Besides, we have to anticipate what the true natural laws are in relation to this complete description, and not the incomplete description. (Naturally this complete description would not be limited to the fundamental concepts used in point mechanics.)
>
> I have told you more than once that I am a fierce partisan not of differential equations, but of the principle of general relativity, whose heuristic force is indispensable to us. Yet, in spite of much research, I have not succeeded at satisfying the principle of general relativity otherwise than using differential equations; perhaps someone will discover another possibility, if they look with enough perseverance.[127]

Thus Einstein had seriously considered the abandonment of differentiability. But this abandonment could at first only be associated with giving up differential equations, then the main mathematical tool of the physicist. This dilemma has been resolved by the introduction of fractal geometries. As we shall see, the scale relativity method, in which one includes scale resolutions in an explicit way in the equations of physics, allows one to describe nondifferentiable phenomena in terms of differential equations (but acting both in spacetime and in scale space).

Feynman and the Return to a Spatiotemporal Representation

Around the same time, Richard Feynman established the connection between nondifferentiability and quantum mechanics. In the 1940s, he tried to return to a spacetime approach to quantum mechanics. Part of this work was published twenty years later in the book he co-authored with Albert Hibbs.

Feynman first developed an approach, the "path integral," based on an initial remark by Dirac. Contrary to the Copenhagen interpretation of quantum mechanics, which completely abandoned the notion of trajectory, Dirac reintroduced the notion of paths for particles. His method allowed for all possible paths (in infinite number) going from one point to another to be taken into account *a priori*. Paths which greatly diverged from the classical trajectory cancelled each other out by destructive interference. On the other hand, paths which stayed close to the classical trajectory (in an area which could be quite large) became highly probable due to constructive interference.

Feynman then pushed this analysis further, to study the typical paths of a particle in quantum mechanics, that is, those which make the largest contribution to the final probability distribution. Can we characterize them by some interesting property?

Feynman writes:

The important paths for a quantum-mechanical particle are not those which have a definite slope (or velocity) everywhere, but are instead quite irregular on a very fine scale. . . . Thus, although a mean velocity can be defined, no mean-square velocity exists at any point. In other words, the paths are nondifferentiable. . . . The square of the velocity is of the order $1/\delta t$ and thus becomes infinite as δt approaches zero.[128]

What Feynman describes here (and which he had understood by the 1940s) is exactly what we now call a fractal curve (a concept which did not exist at the time). Better yet, his discovery that the square of the velocity varies as the inverse of the time interval is equivalent to the statement that it has a fractal dimension of two. This result was obtained using different methods by Abbott and Wise and several other authors at the beginning of the 1980s. These potential quantum trajectories, while they are all different and infinite in number, all possess a common geometric property: they are all fractal curves of fractal dimension 2.

Thus Feynman not only established for the first time the nondifferentiability of quantum paths, but went further and described them in terms which are identical to those which Mandelbrot later named fractal curves. If Feynman had known this concept and that of their connected fractal dimension, he might well have expressed his results in these terms and further pushed the geometric interpretation and analysis.

But even without Feynman, the Copenhagen interpretation of quantum mechanics itself says nothing else: the abandonment of the concept of trajectory precisely means the abandonment, at least in part, of the *classical* concepts of acceleration, velocity, and eventually position. In other words, the abandonment of the concepts of the differentiability of velocities and coordinates, perhaps even of continuity.[129] The Heisenberg uncertainty principle itself removes all physical meaning of the limit "δx or δt tending to zero," since one must use an infinite momentum and energy to make such an infinitely precise measurement.

I must insist here on one point: to abandon differentiability does not mean abandoning *continuity*. Mathematicians have known since the nineteenth century that there exist continuous but nondifferentiable functions (which appeared, at the moment of their discovery, to be kinds of "monsters.")

Since Newton, all of mathematical physics is based on integration and differentiation. The apogee of classical physics is the theory of general relativity, of which the equations are invariant under *continuous and twice differentiable transformations* of the coordinate system, as we have recalled. It is thus clear that a future physics allowing for processes everywhere nondifferentiable could in no way be classical, since classical physics rests in its entirety on this axiom, even if it is often implicitly so.

An Extension of Relativity?

A geometric approach to the quantum problem thus seems possible, an approach which would consist of generalizing the principles of relativity and of covariance to changes of coordinate systems which are no longer *a priori* differentiable. What might be the equations of such a "super-relativity," invariant under these new extremely general transformations? They would automatically include Einstein's equations in generalizing them, and would thus describe new structures going beyond gravitation. What would these new structures be? Would it be possible that they include quantum laws?

Trying to resolve this problem in this form proves to be an extremely difficult project. This is for a very simple reason: differential calculus is the fundamental tool of physics since Leibniz and Newton. The laws of modern physics, as Poincaré so well expressed, are those of differential equations. If one were to abandon differentiability, would one not also abandon differential equations, as Einstein feared? This would be to abandon the same method which has made physics an "exact" predictive science. What tool should be used instead? Everything would have to be redone.

Fortunately, another way is possible which, in an astonishing manner, boils down to all the preceding in providing a mathematical tool which brings a solution to the problem I have just raised. Before such a wall it is useless to push forward with a lowered head. One must go around it. Such a problem is not resolved, it is surpassed. Scale relativity is precisely the art of dealing with nondifferentiability with the help of differential equations.

Chapter 15
Scale Relativity

The key to the problem can be found implicitly contained in the work of Feynman and in its reinterpretation in terms of fractals. It is clearly not possible to keep the concept of deterministic trajectory for a quantum particle, since the number of possible paths is infinite. Nor is it possible to keep the concept of velocity in the classical sense over any of this infinite number of potential paths. This velocity being formally infinite, Feynman tells us that it "does not exist." But it is only infinite *at the limit* where we make the interval of time tend to zero.

Here we have put our finger on the essence of the problem! The differential method in physics is based on the Cartesian project, which could have been called reductionist (but to which we owe three centuries of extraordinary successes, one must not forget): one decomposes the complex object to study it in its smaller parts. This simplicity allows a local description (differential) which, after integration, allows one to grasp the global properties of the system.

But what happens if the parts, instead of being "more simple," are shown to be more complex, or simply different? If, in observing the object under consideration with a microscope (or at a smaller scale, with a particle accelerator), instead of observing the expected smooth, "rectified" behavior, new structures appeared, then others, then others again, and so forth without end, which is just a general expression of fractality? This continual appearance of new structures over the course of a "zoom" toward smaller and smaller scales is precisely the result obtained in experiments. It is true of the quantum "objects." Could it not also be true of the spacetime which contains them?

The Structure of the Electron

The electron itself provides a typical example of such behavior. Although it is authentically an elementary particle, the electron is, from a certain point of view, simpler at greater scale than at smaller scale. In an experiment done under classical conditions, it behaves as a point-like particle, without internal structure. The same is true for the quantum mechanical electron within an atom, at the level of the angstrom.[130] This scale stays "non-relativist," in the sense in which the characteristic velocities involved are of the order of a hundredth of the speed of light.

But if we want to describe the electron at scales one hundred times smaller, lower than its Compton wavelength, it becomes more complex and is subject to curious phenomena.[131] The reason is that the Compton wavelength of a particle is given, up to a constant, by the inverse of its mass. Therefore, if one wants to make measurements at this scale, an energy greater than the rest energy of the particle should be used, which becomes sufficient for creating other, new particles. As a consequence, the smaller the scale at which one "looks" at the electron, the more it becomes complex and made of an increasing number of different particles.

The very nature of electric charge leads the electron to emit and absorb photons continuously, the particles of the electromagnetic field. Indeed, the charge is nothing else than this capacity of emitting and absorbing photons. Some of these photons can be emitted by an electron and be absorbed by another particle (which creates an interaction between the electron and this particle). But the majority of them are reabsorbed by the electron itself. These photons are called virtual, since one cannot observe them directly. To observe them would be to capture them by another particle, before their absorption by the electron that emitted them. Their lifetime, given by the Heisenberg relation, is extremely short: the more energetic they are, the shorter they "live." They are thus, in a certain matter, part of the electron. They also contribute to its "self-energy," that is, to its mass.

But over scales of very short length and time, these photons become so energetic (above two times the mass of the electron) that they can be transformed for a very brief time into an electron-positron pair (see figure 18). At scales of length two hundred times smaller, (thus at energies two hundred times greater), one crosses the threshold of the creation of muon-antimuon pairs.[132] And, gradually, in diminishing the interval with which one considers the electron, one encounters at its interior the whole set of elementary particles, in forms of particle-antiparticle pairs. These include, among others, in addition to the electron and the muon, the tau lepton and the up, down, charm, strange, bottom, and top quarks.

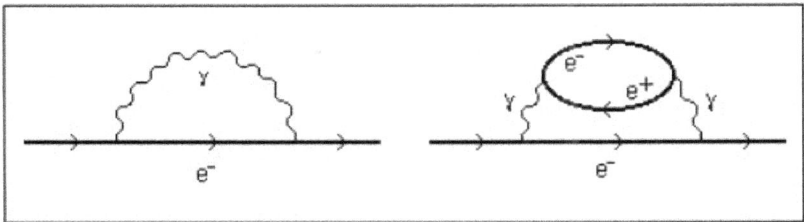

Figure 18 *Virtual particles internal to the electron.* An electron perpetually emits and absorbs photons (which is the nature of its electric charge). Over very short intervals of time, these photons exceed by two times the energy of the mass of the electron and can be transformed in an electron-positron pair (right part of the figure).

These internal structures are called virtual due to their brief lifespan, but they are effectively real since their existence determines the production of the mass and charge of the electron.[133] In fact, the calculation of their contribution to this mass and this charge even gives them infinite values within the standard quantum framework! These infinities are not directly observable since they correspond to the "naked" electron, seen at the smallest possible scale. On the other hand, the contributions of the cloud of virtual particles which the electron "wears" imply that its charge and its mass should vary as a function of the scale beneath the Compton wavelength (see figure 19).

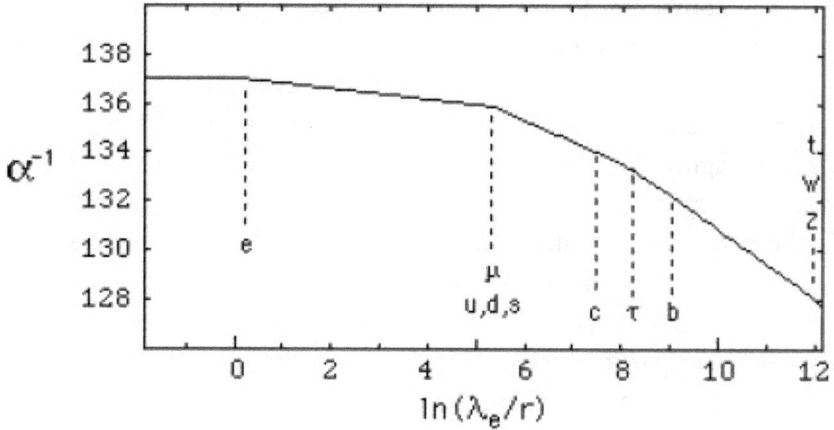

Figure 19 *Variation of the electric charge as a function of the scale of length.* I have plotted here the inverse of the square of the charge (which one calls the "fine structure constant"), in dimensionless units. The distance from one tick to the next along the horizontal axis corresponds to a diminution of the considered scale by a factor of 2.72. The charge stays constant and equal to its macroscopic value (≈ 137) up until the Compton scale of the electron. Smaller than that, virtual electron-positron pairs contribute to its value, which increases at lower scales. For intervals of length that are 200 times smaller, pairs of muons and then of quarks contribute as well, which amplifies the variation of the charge.

This variation has been experimentally observed: while the square of the electric charge is only equal to 1/137 at the Compton scale of the electron, it is equal to 1/129 at the scale of bosons of weak interaction, 200,000 times smaller. Thus the electron can be considered as simple at large scale and more and more complex at smaller scales.

Resolutions in Physics

The measurement of spatiotemporal coordinates is always performed at a certain finite *resolution*. It is thus with whatever kind of measurement in physics.

In certain cases the resolution corresponds to the precision of the measuring device: thus, in the domain of classical physics, a measurement made with a better resolution will give the same result with greater precision. In other cases, it corresponds to the existence of an intrinsic physical limitation. For example, it is doubtful that the distance between the Earth and the Sun could be measured at a precise instant with an uncertainty of one fermi.[134] Such a measurement is expected to never have any physical meaning, since it would require determining the positions of centers of gravity of two macroscopic bodies with the resolution of the nucleus of an atom.[135]

But in the case of quantum physics, the physical status of resolutions changes in a yet more radical manner. The resolution of the measuring device plays a completely new role with respect to the classical case: the results of measurement depend on it in an essential manner, as described by the Heisenberg relations. These relations state that the result of a measurement of momentum depends on the spatial resolution at which it is made: the fluctuation of values obtained for momentum will be inversely proportional to the spatial resolution. Similarly, the result of a measurement of energy depends on the time resolution.

It is not useless to insist on the importance of resolution in physics. A given set of data has no meaning unless it is accompanied by "errors of measurement" or "uncertainties," and more generally by the resolutions characterizing the considered system. Complete information about the measurement of position and instant are only obtained if one gives not only the spatiotemporal coordinates, but also the resolutions with which these coordinates are measured. While such an analysis already plays a primary role in the theory of measurement and interpretation of quantum mechanics, one must still observe that its consequences concerning the nature of spacetime at quantum scales have not yet been drawn. This lack could come from the fact that the resolutions do not appear in an explicit manner either in the definition of coordinate systems or in the fundamental equations of physics. This seems to contradict the fact that they become essential variables in quantum theory, since they carry a part of the information needed to understand the physical significance of the results of measurement. As we shall see, one of the main goals of the theory of scale relativity is to solve this problem.

Relativity of Scales

Intervals of length and of time are always relative quantities: no absolute scale exists in nature (even if universal scales exist). Thus, just as there is relativity of positions and of instants, of orientation of axes and of motion, there is also relativity of scales. The resolutions with which positions and instants are measured, being themselves given by intervals of length or of time, are subject to this relativity of scales. One of the essential propositions of the theory of scale relativity is that spatiotemporal resolutions must be included in the very definition of coordinate systems. Numerous arguments lead to this new conception as we shall now see.

Changes of Units

Up until now, physics has considered transformations of spatio-temporal coordinate systems which correspond to changes of the origin and orientation of axes (changes of orientation in space-time includes movement in space). The relative origin of the coordinates, the relative angles between the axes, and its relative velocity are the quantities which characterize the state of the coordinate system.

Do there not exist other essential magnitudes which could play a similar role?

The results of measurement in physics depend, in particular, on the choice of *units*. Several attempts have been made at including the transformation of units into physical laws. However, the choice of unit is in most cases a purely arbitrary choice which does not describe the conditions of the measurement, but only their recording in terms of numerical result. One does not expect to see the change of units play anything but a trivial role.

Nevertheless, we can still look further into the meaning and the physical role of units. There is a radical way to realize the incompleteness of our present definition of a coordinate system: it is constructed by drawing axes (3 for space or 4 for spacetime) and defined by giving the (relative) origin of these axes, their (relative) orientation, and their (relative) motion (velocity, acceleration). This is the current definition. But try to make a measurement with such a coordinate system: it is impossible. Measurements can be done only if you draw ticks on the axes (according to units and the measurement resolution), but this is not explicitly specified in the used definition. As a consequence, changing the state of the reference system with this current definition includes changing position, orientation, and motion, but not resolution. A complete description of reference systems should therefore include the measurement resolution in the variables that characterize their state.

The necessity of using units to measure intervals of length and of time is directly tied to the relativity of all scales in nature. When one claims that one is measuring a length, one effectively determines the ratio between the lengths of two bodies (one of which is taken as the unit). The speed of a body has no absolute physical meaning, but only has meaning as the speed of a body relative to another, as Galileo discovered. Motion is not something intrinsic to a body, there is no velocity of an object, but only an inter-velocity between two objects. In the same way, the length of an object or the period of a clock has no physical meaning in itself, but only has meaning as the ratio between the lengths of *two bodies* (the object and the body that serves as unit) or the ratio between the periods of *two clocks* (the temporal phenomenon being measured and the clock that serves as reference).

When one says that a body has a length of 132 cm, what one means is that a second body, to which we arbitrarily assign a length of 1 cm and that one defines as a "unit," must be dilated 132 times to obtain the length of the first body. The *direct* measurements of intervals of length and of time always boil down to, ultimately, ratios of dilation. The tendency for physicists to define a unique system of units has certainly been a good thing, which has allowed a rational comparison of measurement results between different laboratories and countries. However, the fact of relating all lengths to a single reference body (or a single period for all measurements of time) has given the false impression of absolutism: this method masks the essential characteristic of the relation between intervals of lengths, which is in fact a relation taken two at a time.

What has allowed this use of a single unit is the simplicity of the law of the composition of dilations, given directly by their product. There is no doubt that this law is extremely well verified in the classical domain: a body of length 2.1 m measures 210 cm as well with certitude, knowing that one meter is equal to one hundred centimeters. However, one can observe that the knowledge we have about laws of dilation in the two domains of quantum physics and of cosmology is only indirect. Metric kinds of measurement of length and of time are impossible in these two domains. The values assigned to intervals of length and of time here are deduced from observations of other variables (for example energy and momentum of particles at small scale, luminosity and apparent diameter of galaxies at large scale) and from an underlying theory (respectively, of quantum mechanics and of general relativity), which have been constructed under the implicit hypothesis that the usual laws of dilation are correct. Such a situation can be compared to the status of the laws of motion before the advent of special relativity: it seemed equally obvious that the law of composition of velocities had to be their direct sum, $w = u + v$. But Poincaré and Einstein have demonstrated that this law is valid only for small velocities and should be replaced by a more general one.[136] We will see that such an analysis effectively leads us to make new propositions concerning the nature of the laws of scale in the two asymptotic domains of very large and very small scales.

Change of Resolution

Let us now consider the physical status of *resolutions*. This status is deeply connected to that of units: the interval of resolution with which a measurement is performed corresponds to the minimal physical unit accessible for the experiment under consideration. There is no change of physical conditions if one decides to express a result of length measurement in another unit, for example (8.152 ± 0.001) m in the form (815.2 ± 0.1) cm; the limit of such a reformulation is attained when the chosen unit becomes of the order of the resolution (which is 1 mm in this example): this would yield (8,152 ± 1) mm. The use of a unit much smaller than the millimeter to express such a result would make no sense. For example, if one were to obtain a result like (8,152.003 ± 1) mm, the last digits would have no physical meaning. On the other hand, reperforming the measurement of the same length with a higher resolution takes a new and essential physical significance: the conditions of the measurement have changed, not only the expression of the result. For example, at the resolution of 1 μm, one could find 8,152.076 ± 0.001 mm, a result which would now be meaningful and in agreement with the measurement at a much lower resolution. But the new digits include much more information.

A change of resolution corresponds to an explicit change of experimental conditions. To measure a length with a resolution of 1/10 mm requires the use of a magnifying glass; at 10 μm, one needs a microscope; at 0.1 μm, an electron microscope; at 1 Å, a scanning tunneling or field-emission microscope. At yet smaller scales, measurements of length become indirect, knowing that we have attained, and then passed beneath the size of atoms.

When one enters into the quantum domain, that is, for resolutions lower than the de Broglie lengths and times of a system (this will be clarified later), the physical status of resolutions changes radically, as I have already shown.[137] While in the classical domain one can equate resolution with the precision of the measurement (two measurements performed at different resolutions produce the same result, but with more or less precision), in the quantum domain, on the other hand, the results of the measurement are affected by the resolution of the device.

In quantum mechanics, the dependence on scale is already implicitly present, in particular with respect to the Heisenberg uncertainty principle. But it is explicitly apparent neither in the axioms, nor in the physical variables, nor in the equations. Under these conditions, should current theory not be considered as conceptually incomplete? Should a complete physical theory not include in its variables and in its equations the entirety of physical information provided by the experiment? In other words, it is a question of asking that the theory of measurement, rather than being added in an external manner to theory (under the form of interpretive statements posed *a posteriori*), should be an integral part of it, and that the essential observed dependence for physical laws as a function of spatiotemporal resolutions be expressed right at the level of the fundamental equations of physics. Such a theory should not need interpretation if the operating rules are posed beforehand at the level of its foundations.

There is a more general argument pointing out the need for an explicit introduction of scale variables in physics. If one asks a student in physics to write the equation of motion of an object, they will naturally write Newton's equation of dynamics. But if one specifies that the object size is actually smaller than one Angstrom, the answer is that one should use the Schrödinger equation. In scale relativity, as we shall see, there is a spontaneous and natural transition from one equation to the other, instead of just a "diktat" as in present physics.

Universality of the Heisenberg Relation

The interpretation of quantum mechanics can then be more deeply analyzed by the light of these ideas. I have made the observation that, among the properties of physical objects, only those that have a character of universality can be compared to the properties of spacetime itself. In the quantum domain, this criterion of universality leads us to two fundamental relations in particular: that of de Broglie and that of Heisenberg.

The universality of the Einstein-de Broglie relation is a manifestation of the universality of wave-particle duality: all physical systems, not only elementary ones, possess wavelike properties characterized by the de Broglie period and wavelength, which are inversely proportional to the momentum and energy of the body. This has been verified by experiments involving interference and diffraction of elementary particles like electrons or photons, but also on composite objects like nuclei or atoms.

As for the Heisenberg relation (recall that it states the existence of a minimal possible value for the product of resolutions in position and in momentum), it is a direct consequence of the formalism at the base of quantum mechanics. Even if the Heisenberg relation is an inequality and not a strict equality, it is a universal law of nature. However, this law, in spite of its universality, is often considered in current quantum theory as a property of quantum "objects" themselves, or rather of the mechanism of measurement.[138] One might object to this point of view that the Heisenberg relation can be established in a general manner without alluding to any particular measurement, knowing that it is deduced from a purely mathematical property. It is thus a direct consequence of the wavelike nature of quantum systems. The difference with the lower limit value (which is a fraction of the reduced Planck constant $\hbar$)[139] fluctuates depending on experimental conditions, but it is not so for this limit itself which is completely independent of the measuring device.

A new interpretation of quantum measurements is then possible: the essential dependence of physical laws as a function of spatiotemporal resolutions, which are manifest in the Heisenberg relations, could preexist all measurements and constitute a geometric property of spacetime itself. The effective measurements will do nothing other than reveal this universal property of nature. Thus, one way to implement the principle of scale relativity would be the introduction of a new spacetime possessing such properties of universal dependence on scale.

It is remarkable that we have arrived at the same conclusions as before, but via a completely different path. We had started from an analysis of the current limitations of the principle of relativity: being limited to changes of twice differentiable coordinates, one feels the necessity of its extension to nondifferentiable motion. This is especially true after Feynman's work on the continuous but nondifferentiable character of the typical paths of particles in quantum mechanics. Yet one of the key principles of the entire scale-relativistic approach is the theorem according to which the continuity and the nondifferentiability of spacetime require its explicit dependence on resolutions. We had previously seen that an attempt of geometric translation of Heisenberg relations leads us to introduce an explicitly scale-dependent, i.e. fractal, spacetime. Now, following a different and independent avenue, that of generalizing the differentiable spacetime of Einstein's relativity to a continuous but nondifferentiable geometry, we also arrive at the same concept: namely, a spacetime which would be fundamentally dependent on scale, in such a way that lengths measured inside it become infinite when the scale interval tends to zero. This means, following the general definition adopted here, introducing the notion of *fractal* spacetime.

The combining of the two desired generalizations of relativity, that of nondifferentiable motion and that of scales, is now complete. It is not a matter of adding another hypothesis to physics (that spacetime is fractal in nature at the microscopic scale), but on the contrary of going toward a greater generality, and thus toward a yet more inevitable version of physical laws. One *abandons* the implicit hypothesis that spacetime is differentiable, which, in maintaining its continuity,[140] *necessitates* its explicit dependence as a function of resolutions and its divergence toward small scales. But where it is a matter of relativity and of covariance, and not of an arbitrary generalization without constraint, we must require that the new laws written for such a nondifferentiable (and hence fractal) spacetime keep the same form as in the differentiable case.

The Principle of Scale Relativity

We have at last arrived at the definition and statement of the principle of scale relativity, the principle which will allow us to construct the new scale laws. Indeed, such an approach implies that the various physical variables, in particular the coordinates themselves, become explicitly dependent on scales of spatiotemporal resolution. This explicit new dependence is expected to be described by well-defined physical laws of scale. How can we construct such new laws, which are absent from standard physics? For that, we need a fundamental principle. The history of physics has shown that the most deep and efficient principle upon which to construct the laws of nature is the principle of relativity itself. It therefore seems natural to derive the new scale laws from the principle of relativity, once it has itself been extended to scale transformation of the coordinate systems.

The idea which underlies scale relativity consists of replacing the usual physical quantities with functions explicitly dependent on resolutions. We can see why by considering the first of these quantities, spacetime coordinates. The coordinate of a point situated on one of the axes of a reference system (eventually curvilinear) is the distance to the origin measured along this axis. This coordinate, once the units have been chosen, is a number. But suppose now that we define a system of axes which are no longer curvilinear, but fractal. This can be a choice (but a complicated one!) in a flat or curved space. On the other hand, in a fractal space, it is a necessity; we have no choice since all coordinate systems in the fractal space are themselves fractal. In this case, with the usual methods, which amount to placing oneself at the limit of zero space and time intervals, the coordinate is the length of a fractal curve measured between two distant points (one point is the origin of the reference system, the other is the point itself of which we measure the coordinate): it is thus infinite. The problem seems to be unsolvable (like "attempting to breathe in empty space," said Einstein).

But if one now explicitly introduces resolution, this "length" returns to being finite at all non-zero resolution intervals. Considering all possible resolutions, the coordinate is no longer a number, but an explicit function of the resolution. This function can be clearly defined and known: the apparent impossibility of working with an infinite quantity is solved in this way, being translated by the fact that this function just tends to the infinite when the resolution interval tends to zero.[141] It would be the same for all physical quantities depending on spatiotemporal coordinates.

The resolutions then take on the new meaning of essential variables, intrinsic to the nature of spacetime. Our task now is to consider the *continuum of all possible resolutions* and the relations which tie them together, in the form of a "scale space."[142]

All these elements can now be combined to arrive at the proposition, made independently by Garnet Ord of Ryerson University and myself at the beginning of the 1980s, that quantum properties are a result of the fractal nature of spacetime at small scales.

However, once resolutions have been explicitly introduced, what physical meaning do we give to them? A first possibility consists of extending the notion of reference system by introducing spatiotemporal resolutions as new coordinates. One can imagine the introduction of resolutions as the attribution of a *thickness* to the axes of coordinate systems: this corresponds more to the reality of measurements actually performed, involving the impossibility of actually making use of the infinitely fine axes that we implicitly assume in physics today.

Nevertheless, this concept of resolutions as generalized coordinates is found to be insufficient. It does not incorporate the analysis made above on the relative character of resolutions. The fact that spatiotemporal resolutions possess this same property of relativity which already characterizes motion is nevertheless remarkable: *only a ratio between two scales has meaning; one cannot define an interval of length or of time in an absolute manner.*

We will thus interpret resolutions not as new coordinates of the reference system, but as *physical quantities characterizing its state.* Similar to how velocities characterize the state of relative motion of the coordinate system, spatiotemporal resolutions will define its *scale state,* a state of scale that is always relative to another system of reference. This redefinition, by assuring a complete parallelism with already constructed theories of relativity (which concern static displacements and motion), allows us to apply relativity to scales themselves.

In addition, by extending Einstein's formulation of the principle of relativity of motion, a *principle of scale relativity* can be stated in the following form:

> The laws of nature must be valid in all coordinate systems, no matter what their scale state.

The extended principle of relativity will require the validity of the laws of nature in all coordinate systems, no matter what their state of motion *and of scale*. The mathematical translation of scale relativity will finally be scale covariance (which will complete the covariance of motion):

> The equations of physics keep their form (are covariant) under all transformations of scale (that is, under the contractions and dilations of spatiotemporal resolutions).

Chapter 16
Special Scale Relativity

The power of the principle of relativity, in particular its constructive character which we insisted upon earlier, can now be used to establish the structure of the new scale laws. What are the laws of dilation and of contraction of resolutions? What are the laws of the dependence of various physical quantities as a function of the variation of scale which come under the principle of relativity? This question can take two forms: are the laws currently accepted, deduced from experiment (such as those of the composition of dilations, given by their direct product, or the power laws of constant exponents of ordinary fractals) in agreement with this principle? Are they the most general laws possible?

In other words, are the laws which seem "simplest" to us indeed those which are optimized by nature? What general laws are compatible with the principle of scale relativity, even in the more restricted case of linear scale transformations?

The answer to this simple question, as we will now see, leads to the proposition of a profound paradigm shift for the physics of the infinitely small and the infinitely large. We will follow slightly more technical developments in the two next sections to show how we will answer the question. But it can be summarized here in a few words: similar to how the relativity of motion leads us to introduce an unsurpassable speed (the velocity of light), scale relativity concludes that there exist two finite, unsurpassable scales, one minimal and the other maximal, possessing the physical properties which have heretofore been given to zero and infinity.[143]

"Galilean" Scale Laws

The first point to examine concerns the problem of laws of ordinary contraction and dilation, which are generally considered to be beyond dispute—but we must recall how before the advent of the special relativity of motion, the law of composition of velocities given by their simple sum seemed equally unassailable.

If one starts with a scale λ_1 and one dilates it with a factor ρ, one obtains a new scale $\lambda_2 = \rho\lambda_1$. A new dilation of λ_2 by a factor ρ' gives us a scale $\lambda_3 = \rho'\lambda_2$.[144] In contemporary physics, one supposes that the dilation ratio between λ_3 and λ_1 is $\rho'' = \rho\rho'$ and is thus given by the product of the two initial dilations. For example, it is assumed without debate that if a scale (#1) is two times greater than another (#2), itself three times larger than a scale (#3), the first (#1) is six times larger than the third (#3).

By using a logarithmic notation, scale transformations become most meaningful. This notation is naturally introduced when we are dealing with scale ratios. For example, a zoom lens on a camera will tend not to enlarge by 2, then 3, 4, 5, 6 times, but instead 2, 4, 8, 16, 32 times; measurements of sonic intensity are made in decibels, also defined in terms of logarithms. This transforms the product of dilations to a sum. A dilation by a factor $10 = 10^1$, then by a new factor $100 = 10^2$, leads to a factor of $1,000 = 10^3$. In other words, logarithms transform the product $10 \times 100 = 1,000$ into the sum $1 + 2 = 3$. In this natural form, the law of composition of dilations takes the same form as that of velocities in Galilean theory (a sum).

The comparison does not end there. Let us measure the length of a standard fractal curve (of constant fractal dimension) between two points, at a given resolution, and then at another resolution. How does the length of the curve change over the course of the change in resolution? By analyzing the effects of such a transformation, we shall be able to put to work the principle of scale relativity, similar to how the laws of relativity of motion can be established by studying the effects of a change in speed.

The measured length along the fractal curve is considered, in the framework of scale-relativistic theory, as a generalized curvilinear coordinate: we assume we can work on a fractal coordinate system. As for the change in resolution, it is considered as a change in coordinate systems: but it is now the scale state of this system that has changed, when the origin, the orientation of axes, and the state of motion have remained the same. The invariant of the transformation is the fractal dimension, which is assumed to stay constant.

The result we obtain is fascinating: one can show that a scale transformation on such a fractal coordinate takes precisely the mathematical form of the Galilean transformation of motion![145] The length here plays the role of the coordinate of position, the fractal dimension the role of time, and the resolution, as it should, of velocity (length and resolutions being taken as logarithms, as expected for scale laws). Self-similar fractal behavior (see figure 4) is thus the simplest solution to the problem of scale transformations and is revealed to be the equivalent for the scale laws of what uniform rectilinear motion at constant velocity is for the laws of motion.

In other words, self-similarity is a sort of "scale inertia." The analysis by relativity had allowed the conclusion that Galilean motion "is like nothing," and that it demonstrates the simplest of possible motions, without necessary cause. One arrives at the same type of conclusion concerning self-similar fractals (of constant fractal dimension). Once one allows that dependence on scale is intrinsic to the nature of things, since it just manifests a greater generality (the abandonment of the differentiability hypothesis), one can look for the simplest possible scale laws, which are shown to be those of self-similar fractals. This is why *"fractals are everywhere."*[146]

We can now answer the first question which we asked: the scale-invariant fractal laws satisfy the principle of relativity, since the Galilean transformation is indeed a solution (even if it is particular and degenerate) of the relativity problem.

One can go still further. In effect, it is not a case of a simple analogy between laws of motion and laws of scale, but indeed the same fundamental problem. In the two cases, movement (in one dimension) and scale, we look for the law of transformation of two variables, depending on a relative parameter (respectively velocity and resolution) which comes under the principle of relativity. Should the same problem not have the same solution?

Yet the Galilean solution, as we now know in the case of the laws of motion, is only an approximation, the correct solution being the Lorentz transformation. The Galilean transformation is a very special case of the Lorentz transformation, in which the speed limit c is infinite, which causes a sort of degeneration of the law. What about the scale laws?

"Lorentzian" Scale Laws

To answer such a question, one must forget what one knows (or what one thought one knew) about the laws of dilation and contraction, and ask the problem without any *a priori* information. It is a matter of determining how two variables, length and fractal dimension, are affected in a transformation which depends on only one parameter, the resolution. This will oblige us to a new attempt at abstraction. The search for a more general law leads us to introduce a "fractal dimension," which itself becomes variable. It could then play a totally new role for the scale laws, that of a dynamic variable.

To better understand this point of view, one must remember that what one now calls time in physics has undergone a similar evolution. Before Galileo, time was "the measure of motion." It was not seen as a primary variable, but as something to be deduced from positions and motion. Galileo invented spacetime (even if space and time remained decoupled) by treating the time and space variables on the same plane, and by deducing from them the notion of velocity. It is a similar evolution that the relativity of scales suggests. The variables of position (on a generalized fractal curve) and the fractal dimension are now treated at the same level, and the spatiotemporal resolutions become *derivative* variables (similar to how velocity is the derivative of position).

Like in the relativity of motion, the difficulty of the general problem (*a priori* highly nonlinear) leads us to consider provisionally the special problem of *linear* transformations (this is "special" scale relativity).[147]

This problem is posed for the scale laws in the same mathematical form as in the case of motion. The general solution to this problem is well known: it is not the Galileo group, but the Lorentz group, as we have previously recalled. The way by which this fundamental result is obtained is similar for motion and for scale. In both cases, one first states the principle of relativity in its "philosophical" form ("the fundamental laws of nature are valid whatever the state of the reference system – of motion and of scale"); then one translates it in its physical form of covariance principle ("the equations of physics should keep their form in any change of the reference system – of motion and of scale"); and, finally, it becomes a mathematical statement and proof, by writing in an explicit way the expression of these changes of coordinate system, and by requiring that the invariance of form of the equations apply in particular to the equations of transformation of coordinates themselves. The result of this approach is a proof of the Lorentz transformation as a universal solution of the "special relativity" problem. This proof is based on the mere principle of relativity, and mathematically translated in terms of only two axioms: closure (the first axiom of group theory: the form of the law of transformation by two successive dilations is the same as that of the initial dilations) and reflection invariance (the scale variable being logarithmic, changing its sign means going from a dilation to a contraction by the same factor: the form of the laws of change of coordinate systems should not vary under this transformation).[148]

This reasoning naturally leads to the idea that the standard laws, currently assumed, of dilation and contraction, which are considered as undeniable truth, are only an approximation, valid only at our current scales, of more general laws. The structure of these new laws becomes "Lorentzian" toward very small scales (beyond the Compton scale, meaning very high energies, larger than the mass energy of particles).

Within such laws, the fractal dimension is no longer constant, but becomes variable and plays for scales the same role that time plays for motion. This variable is now combined with the fractal coordinates to form a *vector* of a five-dimensional space: fractal coordinates and fractal dimension no longer are independently transformed, but collectively as *components* of a vector. If one follows the principle of scale relativity, there is really no choice: the general solution to the problem of linear scale transformations is the Lorentz transformation, of which the Galilean transformation is only a degenerate and extremely special case.

There is a third solution to this problem, which is the standard laws of rotation in Euclidean space. Concerning motion, this solution is indeed implemented as projection effects during a rotation, which is in effect a manifestation of the principle of relativity, applied to orientation. But concerning scales, this solution is not applicable, since the product of two dilations could yield a contraction of the initial scale, which can easily be excluded.

The Planck Length as Invariant, Unsurpassable, and Unreachable Scale

What is the nature of the new laws? They are characterized by several original properties compared to the usual scale laws. Let us first consider the new law of composition of dilations and contractions. One can show that it takes the same mathematical form as the law of composition of velocities in Einstein's theory, but these are now the dilation ratios between resolutions which play (via their logarithm) the role of velocities.[149]

The principal new property of this transformation is the appearance of a scale of minimal length Λ_p, which cannot be exceeded (toward smaller scales), and is invariant under dilations and contractions.[150] It is automatically the same for time: there thus appears a minimal scale $T_P = \lambda_P / c$ for all temporal intervals. Indeed, scale relativity is based on the relativity of motion of Poincaré and Einstein and completes it, but certainly does not substitute it: therefore, the fundamental relations between space and time remain.

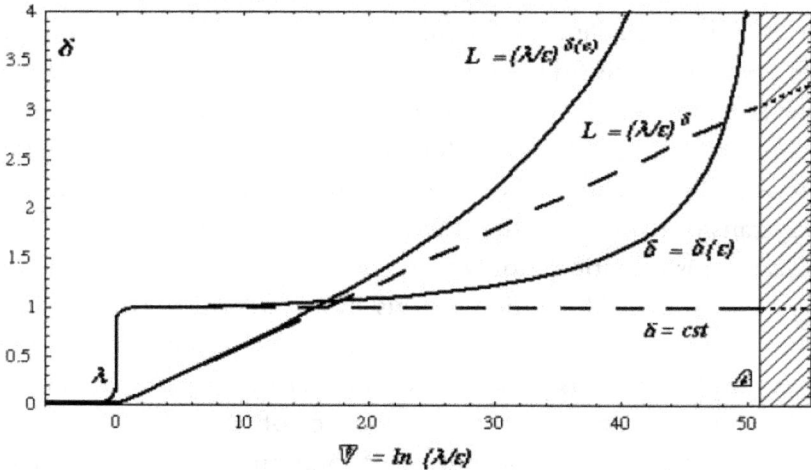

Figure 20 *Fractal laws in special scale relativity* (to be compared with figure 12, knowing that the scales become smaller and smaller toward the right of this diagram). A non-fractal curve at large scales (its length L is constant and its effective fractal dimension D = 1 + δ equals 1) becomes fractal at smaller scales: in the framework of ordinary laws, its fractal dimension D = 1 + δ then becomes equal to 2, and its length diverges according to a power law (dashed line). In the framework of the new Lorentzian laws [solid curve L(ε)], the fractal dimension itself becomes a function of the scale [curve δ = δ(ε)] and tends toward infinity, as does the length of the curve, when the resolution tends toward the minimal, unsurpassable scale (shaded area at right). The numerical values correspond to the case where the scale of transition (V = 0) is the reduced Compton wavelength (3.86 x 10^{-13}m) and the invariant scale limit λ_P is the Planck scale (for which V = ln(λ_e / λ_P) = 51.5). For comparison, the minimal scale attained by current particle accelerators corresponds to V = 12.

The minimal scale will play for resolutions the same role that the speed of light plays for velocities. It replaces the point zero, which no longer has any physical meaning in this framework. But it is important to insist on the fact that it is not a matter of being a wall, nor a barrier or a quantization of spacetime. Rather, the nature of this limiting scale is that of a horizon. It does not put into question the nondifferentiability of the spacetime we are a part of, nor the continued emergence of structures over the course of successive enlargements. It is simply the *effect* of enlargements which is changed. From the point of view of motion, one can indefinitely add velocities to each other (there is no limit to the changes of successive reference systems), but the result will always stay lower than *c*. Similarly, *an arbitrarily large number of successive contractions, applied to whatever initial scale, will lead to a scale which will always be greater than the minimal scale limit λ_P*. Thus, space and time, in the new framework, remain indefinitely divisible. It is the result of these divisions which is limited, not the possibility of dividing.

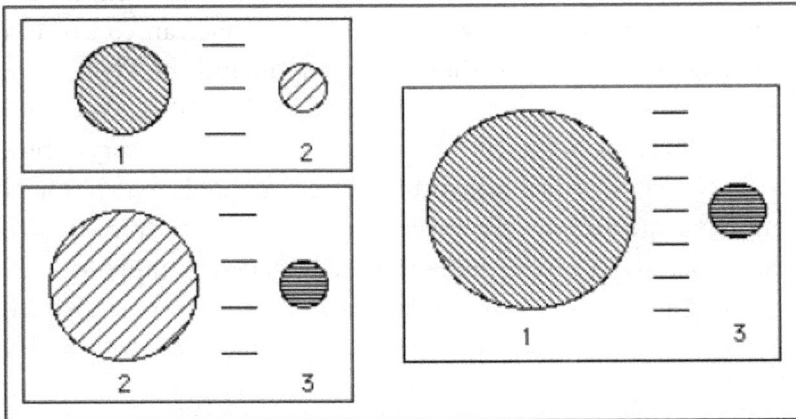

Figure 21 *New dilation laws.* Object 1 is twice as large as object 2. Object 2 is 3 times larger than object 3. Nevertheless, under the new "Lorentzian" dilation laws, object 1 is less than 6 times as large as object 3. All that matters are ratios taken two at a time. The law defining the composition of these ratios is no longer their direct product.

The scale limit in fact possesses all the physical properties once attributed to zero: the scales of energy and momentum now tend toward infinity when one approaches T_P and λ_P, the length of a fractal curve diverges, but also the fractal dimension. All this confirms the unsurpassable character of the minimal length, in the sense in which all scales between it and zero no longer have any existence.

Similar to how we can summarize the relativity of motion by affirming that, from the point of view of velocities, 2 + 2 no longer equals 4, scale relativity proposes that, for resolutions, 2 × 3 would no longer equal 6 (see figure 21).

A final fundamental question is raised: what is the value of this scale limit, unique and universal by nature? Do we need to introduce a new fundamental scale in the laws of physics? Or does this scale already exist in current physics, but simply needs to be interpreted as the invariant scale?

The second proposition seems more natural. In effect, the Planck length scale, $\lambda_P = (G\hbar/c^3)^{1/2}$, seems to have all the required properties to be identified as the scale limit. We have seen that starting from three fundamental constants of physics, G, $\hbar$, and c, one can construct three quantities, the Planck mass, the Planck length, and the Planck time, of which the numerical values—2 x 10^{-5} g (or 1.2 x 10^{19} GeV in energy), 1.6 x 10^{-35} m, and 5.4 x 10^{-44} s—depend on the (arbitrary) choice of our units. They thus constitute three natural units for these fundamental quantities. Moreover, this scale is already recognized in current physics as a sort of natural barrier, since it represents the limit beyond which the effects of gravity become as important as quantum effects. It is the spacetime continuum itself which seems to be broken in the extremely difficult endeavor to construct a theory ("quantum gravity") which would describe the physical phenomena at such energies.[151]

Another argument justifies this reinterpretation of the physical meaning of the Planck scale. Let us consider a thought experiment in which a ruler of length equal to the Planck length would be observed from a reference system in relative motion with respect to the ruler. It should undergo a Lorentz contraction. Yet the Planck scale depends on no particular object in nature. Other fundamental and universal scales exist (like the Compton wavelength of the electron or the electroweak scale), but all are the characteristics of particular physical "objects" (the electron, weak bosons, the Higgs field, etc.). One can thus find it surprising that the Planck scale depends on the motion of the reference system where one measures it. This problem is resolved in the new transformation, since it becomes in this framework completely invariant under all dilations and contractions, *including the Lorentz contraction* of special (motion) relativity.

If the intervention of $\hbar$ and of c in the definition of the minimal microscopic scale seems natural, is it so in the case of G, the characteristic macroscopic constant of the gravitational field? G does not play a part here as the gravitational constant of Newton, but instead as the constant of Einstein, which connects matter and geometry in a universal manner in general relativity. We will see in the following that, in the framework of scale relativity and the fractal approach to quantum mechanics, $\hbar$ also acquires a new meaning as a constant connecting matter and geometry.

Ultimately, if experiment confirms the new role proposed here for the Planck scale, special scale relativity will lead to a refoundation of the question of units of length and time. Let us recall that the special relativity of motion has already led us to eliminate the unit of length. This unit is now deduced from the unit of time and the speed of light, which has been definitively fixed. The speed of light is essentially equal to one. All different values arise from an inadequate choice of units. As soon as one admits the existence of a four-dimensional spacetime and not a separate space and time, the unit of space and the unit of time should be chosen to be identical (in the same way that we use the same unit for measuring widths and heights). Velocities, being the relation of an interval of length over an interval of time, then become quantities without dimension, pure numbers. Finally, special relativity states that these numbers do not vary between zero and infinity, but between zero and one.

It will be the same for intervals of length and of time: if the Lorentzian structure of dilation laws is confirmed at small scale (at the level of the "internal structure" of particles), the Planck spacetime scale will play for them the role of a fundamental "one." In natural Planck units, it would be in the nature of all intervals of length or of time to be a dimensionless number always greater than one.

These new scale laws could have observable consequences at the very high energies attained by large particle accelerators, mainly because of the separation between the Planck mass scale and the Planck spacetime scale.[152] But, before examining some of the possible consequences (we will say more about the nature of the Planck scale in Part Five), we will return to the ordinary dilation laws and show how the concept of fractal spacetime can be applied in quantum theory in the microscopic domain.

Chapter 17
Scale Relativity and Quantum Theory

Introduction

We have seen how an analysis of the properties of the quantum world allows us to form the hope that this behavior, bizarre as it may be, has a geometric origin, but in the framework of a new geometry, nondifferentiable and fractal. Can we go further and attempt to found quantum mechanics on such a basis? In other words, is it possible to reconstruct the axioms of quantum mechanics starting from a first principle, such as that of scale relativity? The origin of quantum behavior would then be clarified as the manifestation of a fractal and nondifferentiable geometry of spacetime at small scale, similar to how gravity and its properties are already understood, in general relativity, as manifestations of the curvature of spacetime at large scale.

We cannot pretend that the theory of scale relativity has completely accomplished such a goal: quantum mechanics retains a large part of its mystery. But several significant advances have been able to have been made which we will now discuss.

The argument can be summarized as follows. The non-differentiability of spacetime implies (at a minimum) three effects. The first is that each geodesic (curve that optimizes the proper time) becomes fractal. The second is that there exist an infinite number of geodesics where classically only one was salient. The third is that the concept of average velocity (which remains definable, as Feynman had emphasized) is no longer single, but double. Taking into consideration these three effects in the equations of motion transforms classical mechanics into a quantum-type mechanics.

Fractal Geodesics

The first step consists of establishing the internal scale laws from which the effects elicited upon motion, one hopes, could be manifested in the form of quantum properties. We have previously seen that fractals of constant dimension appear naturally as the simplest solutions of relativistic scale laws. This is also the behavior obtained by Feynman for potential quantum trajectories: recall that he had found the typical trajectories of particles in quantum mechanics to be (what we now call) fractal curves of dimension 2.

It is within the special framework of a description of geodesics as fractal curves that we will situate ourselves in the following. But one must understand that generalizations of the obtained results are possible, as soon as these restrictive hypotheses are loosened (in progressing to a fractal dimension different from 2, and then a Lorentzian structure of scale laws).

Infinite Geodesics and Probabilities

In a geometric theory of spacetime, the answer to the question—what trajectory do free particles follow?—is a spatiotemporal geodesic, meaning the shortest line at the initial fixed conditions. More precisely, it is the line that optimizes the proper time, i.e. the time given by a clock following this trajectory. In empty, flat Euclidean space, the geodesics are just the straight lines followed at constant speed in Galilean free inertial motion. Einstein understood that the natural generalization of this free motion to different geometries is the concept of geodesic curves. For example, the geodesics on a sphere are the great circles.

In the theory of scale relativity, it is therefore also natural to use this concept for the description of the motion of particles. However, there is a fundamental difference with the case of general relativity. In Einstein's theory, the geodesics are deterministic trajectories followed by material particles. In quantum mechanics, the concept of particle trajectory loses its meaning. As we shall see, this is also the case in a nondifferentiable spacetime. Therefore, it should be made clear that the fractal geodesics are not the "trajectories" of a particle, but are just purely geometric paths from which the wave function (that represents the particle in quantum mechanics) can be constructed. As we shall see, there is an infinity of geodesics in a fractal spacetime which form a fluid, and the wave function comes as a direct manifestation of the velocity field of this fluid.

Indeed, a fractal spacetime is characterized by the presence of obstacles at all scales, which will multiply the number of geodesics to the point of making it infinite. The equation of geodesics of a given space is a differential equation which expresses this optimization of the path followed to an infinitesimal level. One must imagine a fractal spacetime as fluctuating locally, at all scales, in space and in time, in an extremely complex manner (see figures 11 and 22). Therefore, the solution of this optimization is not, in this case, a unique, small elementary displacement, but an infinite number of such infinitesimal intervals!

One can form an idea of this multiplication to infinity of possible paths starting from the effects of gravitational mirage in general relativity.[153] In Einstein's theory, light follows particular geodesics of (4-dimensional) null length. The effect of a mass is to curve spacetime, which implies as a consequence the curvature of geodesics. One of the results of this deviation of light is that, around a point-like mass, two paths can connect a single source to the observer. One then obtains two images of the same object instead of one! Such effects of gravitational mirage have now been observed in great numbers. If the mass is extended, as in the case where the deflector is a galaxy, one finds three or five images. But the distribution of the deflecting mass can become extremely complex, as in the case where it is composed of a cluster of galaxies in its entirety as well as individual galaxies. The number and deformations of images of the same object can then become impressive. And now imagine the effects of a fractal spacetime, in which each point, having an infinite curvature, would become such a deflector.

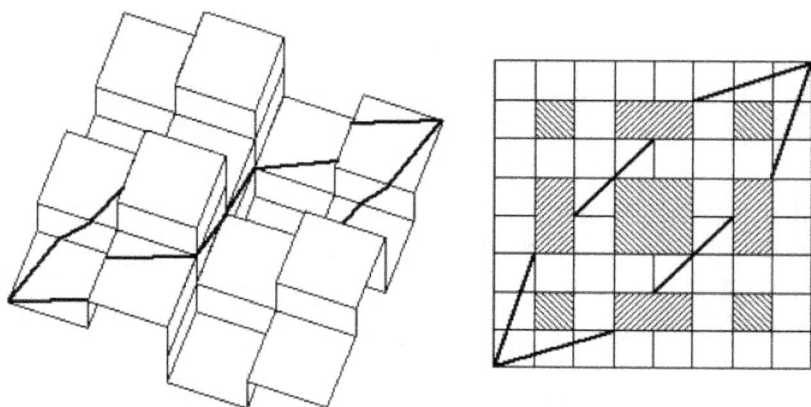

Figure 22 *Geodesics on a generator of fractal surface.* At the first stage of the construction of a fractal surface, even for a very simple and determinate model, the shortest lines between two points (the geodesics) become multiple. (At the left, the continuous folded version, at right, the lacunar unfolded version, in which the shaded squares indicate the points which are connected in the folding). At the following stage, a new multiplication takes place to avoid the new obstacles which appear within each square (see figure 11). At the limit, the number of geodesics is infinite.

As already pointed out, the notion of a particle following a well-defined trajectory cannot remain in such a framework. Suppose that a particle has been emitted at a point. How does one predict its evolution, knowing that it could follow an infinite number of equivalent trajectories? All that we can do now is to describe the evolution of the set made up by all the potential trajectories, which constitute a kind of "geodesics fluid."

The probabilistic approach becomes a necessity as soon as one wants to make predictions. The statistical character of the theory to be constructed is thus established, not as a fundamental law, but as a consequence of the nature of spacetime (and of *our* wish to make theoretical predictions). Spacetime could, in principle, be perfectly determinate,[154] but its fractal nature (in which no lower limit exists to the structures that appear at smaller and smaller scales[155]) implies, in a definitive manner, the nondeterministic nature of trajectories.

Therefore, in the new framework, the quantum indeterminacy is a direct consequence of the nondifferentiable geometry. One sees here how the introduction of a new concept, that of a structured and fractal spacetime, allows us to express in a new manner the problem of nondeterminism. There is indeed nondeterminism, unavoidably so, in the approach of scale relativity, but this nondeterminism is that of "trajectories." It is explained starting from the geometric nature of spacetime, but this spacetime, inversely, is not *a priori* subject to nondeterminism. Such a theory is thus not statistical in essence (at the level of a first principle), as Einstein required in his criticism of the quantum theory, but is of geometric essence and statistical by logical necessity, as a consequence of a deeper principle.

The Nature of Particles

The fractal approach enables us to shine a new light on the concept of the elementary particle. We can envision being able to give up the idea of a mass-possessing point with "internal" properties (spin, charge, other quantum numbers, etc.), and describe the "particles" (with their dual wave-particle nature) as fractal spacetime geodesics. From this perspective, each type of particle would correspond to subsets of geodesics possessing the common geometric properties which define its nature (scale of fractal/non-fractal transition for the mass, internal structure within scales for spin and charge, etc.).

In contemporary quantum theory, the electron, considered from the point of view of its particle-like nature, is totally point-like. The essential physical quantities which define it, such as its mass, its spin, or its charge, are thus considered as purely "internal" quantities. Spin, the internal angular momentum of a particle, is tied to a symmetry of spacetime (isotropy), but has no classical counterpart. Charge and other quantum numbers characterizing the particles correspond to symmetries which are themselves internal, which have no counterpart in spacetime.

The concept of fractal spacetime allows a reconsideration of these statements. It is no longer simply a case of spacetime in the classical sense of the term, structured by movements and rotations (the laws of motion), but of a "spacetime zoom," the description of which necessitates taking into account new structures tied to the transformation of resolution (scale laws). The possibility raised by this observation is that internal quantum numbers finally result from symmetries tied to scale transformations, and that they thus acquire a new geometric meaning, in the sense of the new nondifferentiable geometry. The concept of the particle would no longer then concern an object that "possesses" a mass, a spin, or a charge, but would be identified as the fractal geodesics themselves of a nondifferentiable spacetime, geodesics of which mass, spin, and charge would be a manifestation of common geometric properties.[156] Such a program is naturally far from being completed, but several results pointing in this direction seemed already encouraging at the date of the first publication of this book.[157] Much of these preliminary results have since been confirmed and developed into a full theory.[158]

First of all, the de Broglie wavelength ($\lambda = \hbar/p$) and the de Broglie period ($\tau = \hbar/E$) associated with a particle find a geometric interpretation as a transition from a fractal behavior (at small scale) to non-fractal behavior (at large scales): these are the scales beneath which retrograde motions appear in geodesics, in space and in time. Starting from this de Broglie scale, one can calculate the energy, momentum, classical (average) velocity, the phase velocity, and, finally, the mass of the particle. All these quantities thus find themselves brought back to geometric characteristics of fractal trajectories. Nevertheless, one must understand that it is not a matter of geometric structures in the spacetime (of positions and instants), but instead in the new space of scales, which describes the set of possible resolutions in nature (see figure 3).

A similar result is true for internal angular momentum (spin): in the scale relativity framework, it can be derived and understood in a purely geometric way. The spin cannot exist classically, since it is proportional to the square of the radius of the particle, which is zero. But it is also proportional to the speed of rotation, which can be infinite on a fractal trajectory! The remarkable result is the following: one can show that this product, which is written informally as zero times infinity, is always zero when the fractal dimension of the trajectory is lower than 2 (the classical situation of dimension 1 is a special case), always infinite when it is greater than 2, but becomes finite just for the "critical" value $D = 2$. Yet this value is precisely what one obtains starting from the Heisenberg relation, or in an equivalent manner, what one can deduce from Feynman's analysis of typical trajectories of quantum particles. The existence of spin is then established as a consequence of the fractal geometry. It is remarkable that this kind of simple fractal helix model for spin, first established in the 1980s,[159] has since been fully supported by exact solutions of quantum mechanical equations, themselves exactly derived from the equations of geodesics of a nondifferentiable spacetime.[160]

As for electric charge, one can show that it can be understood as an invariant quantity finding its origin in scale symmetry itself.[161] This new physical description of charge and electromagnetism will be explained in more detail in part V. Since the first publication of the present book, it has been extended to all gauge field theories[162] (which includes also quantum chromodynamics—the strong interaction—and the weak interaction, mixed with electromagnetism in the electroweak field). The various "charges" of these interactions can all be understood in terms of invariants (conservative quantities) constructed from scale symmetries which are specific to the scale space interior of "particles" (i.e., to the inner structures of the fractal geodesics of the nondifferentiable spacetime).

Quantum/Classical Duality

The potential paths thus constitute an infinite set of curves, all of which are themselves fractal. One can show that the description of any one of these curves brings into play coordinates that can be decomposed into the sum of two contributions: a mean part, which is differentiable and scale-independent[163] (and is identical, at the classical limit, with the classical trajectory), and fractal fluctuations, which are explicitly dependent on scale (see figure 23). These fluctuations are negligible at resolutions greater than the de Broglie length of the system: this is classical behavior. But they strongly dominate the average motions at small scale. A large part of quantum effects comes from them.

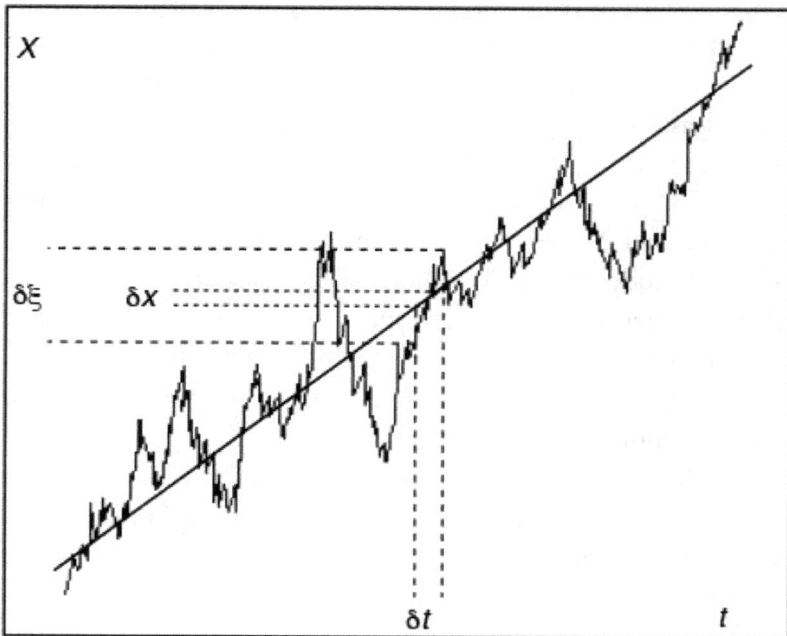

Figure 23 *Evolution over time of a coordinate on a fractal path*. At large scale, there exists an average velocity, and the average variation of the coordinate is $\delta x = <\delta X>$, of the order of δt. On the other hand, at small scale, the fluctuation with respect to this average dominates, $\delta X \approx \delta \xi$, and becomes much larger than δt. It is given in absolute value by $\delta \xi \approx \delta t^{1/D}$, where D is the fractal dimension of the path.

Thus classical and quantum behaviors are a question of scale, but the *relative* character of the transition (which depends on the mass and the velocity, and/or the temperature) explains that there might exist macroscopic quantum effects. It is so, for example, with superconductivity, which appears at low temperature, below a critical temperature depending on the material (in this phenomenon, the free electrons are grouped in pairs and thus acquire the collective quantum properties of a gas of bosons, which enables them to flow without resistance).[164]

Irreversibility and the Complex Character of the Wave Function

The doubling of position variables in terms of differentiable part and fractal part is not all. Another doubling, just as fundamental, comes from the nondifferentiability of spacetime and the principle of microscopic reversibility. It involves not only the fractal fluctuations, but also the mean differentiable part of the velocity.

Consider a potential fractal geodesic arriving at a given point at a given instant. There are an infinite number of geodesics coming out of this point. Each point of spacetime thus plays the role of a sort of diffuser (recall the analogy with a multiplication to infinity of gravitational mirages). But such a process is fundamentally irreversible locally: in a reversal of the sign of the temporal differential element, the nondifferentiability of space implies that there is no longer reversibility of the process.[165] We can be more specific about what this is. From the point of view of usual methods, the derivative of a nondifferentiable variable does not exist (in particular the velocity, which is the derivative of the position), by definition. The method of scale relativity consists of reintroducing a derivative, explicitly dependent on the resolution. It is only at the limit of an interval of zero resolution (the only case considered by classical methods) that the derivative does not exist, whether it becomes infinite or it fluctuates without limit. But this zero limit point does not really make physical sense—I have already heavily emphasized this point—since there would have to be an infinite energy to attain it. One can thus ignore it without any problems, and replace the ideal mathematical derivative with a physical derivative becoming an explicit function of the time interval. But it is here that a new fact appears: while the ordinary derivative is invariant under the change of sign of the temporal differential element, there is no reason that it would be the same for the new derivative.[166] There are thus two derivatives instead of one, each of them being a fractal function! Applied to the spatial coordinate, this reasoning implies that there are two velocities instead of one, a doubling which remains when considering the mean, differentiable part of the velocity.

Such a doubling of the average velocity has already been introduced by Edward Nelson, under the framework of his stochastic approach to quantum mechanics, but with a fundamentally different interpretation. In his stochastic mechanics, one obtains this two-valuedness of velocity by reversing the sign of time itself (which may be problematic with regards to causality), leading to the introduction of a "forward" and a "backward" velocity. In scale relativity, it is only the sign of the time differential element which is reversed: since we need two points to define a velocity (while only one for position), one may take a point "before" or "after" the instant when we compute the velocity. Moreover, Nelson equated the movement of particles to Brownian motion and was able to obtain the Schrödinger equation in terms of a process of nonclassical diffusion. One must nonetheless note that this is only fully justified, in my opinion, in a framework where spacetime itself is affected, not only the trajectories. It is clear that true Brownian motion (that of microparticles immersed in a liquid), while nondifferentiable in its physical description, does not lead to a doubling of average velocities. It is certainly the same if we only consider fractal trajectories in a non-fractal space. Once averaged, these trajectories become classical at large scale. It is the fact of considering, not fractal trajectories *a priori*, but a fractal space of which the geodesics would thereby be fractals, which is here essential. It is because spacetime is arbitrarily chaotic at all scales that the doubling of velocities remains, even for the mean differentiable (large scale) part of the velocities.

Nonetheless, symmetry by reversal of time—that is, reversibility—is a fundamental symmetry in physics, first in relativity, where the "invariance by reflection" plays an essential role in the establishment of the Lorentz transformation, but also in microphysics. Now we have two processes instead of one, which differ under the reflection (i.e., the change of sign) on the time differential element, $dt \rightarrow -dt$. These *two processes should be considered as equally valid* for the description of physical laws.

Here is an essential point. In the classical (differentiable) case, the problem does not come up, since in that case the derivative is unique and does not depend on scale. If spacetime is not differentiable, the two processes become *a priori* different, while they are both equally qualified for the description of elementary laws. Which to choose? One solution alone seems possible: to take into consideration, in a permanent manner, the *two* processes together, in combining them in the form of a new double process. One can prove mathematically that the complex numbers enact the simplest representation of such a double process. This, taken in its entirety, becomes locally reversible. Such is the origin, in this theory of scale relativity, of the complex character of the wave function and other physical quantities, and, in large part, of the paradoxical properties of the quantum world.

One will note that, with this effect of doubling of velocities, nondifferentiability is shown to be more profound than fractality, as already indicated by the fact that scale dependence is a consequence of the abandonment of the differentiability hypothesis (if one keeps continuity).

Before passing to the final step, that of the proof of the Schrödinger equation, the theory should provide a meaning for the wave function. What is the nature of this mathematical tool? Why and how has it come to replace classical tools? The response is simply that, in scale relativity, the wave function is nothing other than a reformulation of the fundamental physical quantity called *action*.[167] This concept is particularly important in physics since the fundamental equations of dynamics are derived from optimizing it (a method called "least action principle"). Moreover, in relativity theories, it becomes proportional to the proper time: therefore, optimizing the action becomes nothing else than optimizing the proper time, which yields just the equation of geodesics. But the new ingredient here is the two-valuedness of derivatives, leading to introduce a twin fluid of geodesics, a complex action (in the sense of complex numbers), and finally equations of motion written in terms of complex numbers.

Another important role of the action is that the energy and momentum of a given system can be obtained by taking its derivative, respectively with respect to time and space. This provides another way to understand the fundamental meaning of the wave function: it just gives the velocity field of the twin fluid of geodesics.[168]

In effect, once a complex derivation is introduced, thus a complex velocity, the entirety of classical mechanics can be reformulated. One must remember that the action is defined as an integral over the course of time of a function of coordinates and velocities, called the Lagrange function. The velocity being now written in terms of complex numbers, it is the same for this function, thus the action and, finally, the wave function itself. The principle of action (which is no longer a principle of least action, since there is no ordering relation in the complex plane) stays definable as a principle of "stationary action": following this principle, the physical paths are those which cancel the variation of action. It is, finally, the entirety of mechanics which can be reformulated by this method, in a way that keeps the *form* of classical mechanics (this is "covariance" under a generalized meaning), but that is no longer classical. The result is indeed the attainment of a new mechanics of quantum nature, as we shall see.[169]

Free Motion and the Schrödinger Equation

The final steps toward this transformation of classical mechanics to quantum mechanics in a nondifferentiable geometry should include the proof of its other main axioms, the principle of correspondence (which makes an operator acting on the wave function correspond to observables), the Schrödinger equation (which is the equation of which the wave function is a solution), and the statistical interpretation of Born (according to which the square of the modulus of the wave function gives the probability of presence of the "particle").

The example of Einstein's theory of general relativity is essential for this final task. The effects of the curvature of spacetime on different physical quantities is there described with the help of a "covariant derivative." This adds new terms with respect to the ordinary derivation, under the form of an effect of rotation of vectors induced by their movements, due precisely to this curvature (meaning the gravitational field). In other words, instead of taking into account the effects of geometry as acting from the exterior on motion (like in the case of forces), one accounts for them *from the interior* by including them in the derivative itself.

However, the covariant *tool* of general relativity, based on differentiability, is no longer adapted to the description of effects of movements in a fractal spacetime. But the *concept* of covariant derivation, which translates these effects in a universal manner (one must construct the rules which make it applicable to all physical quantities), is still fundamental and should be kept and reconstructed. It is effectively possible to summarize the three principal effects of abandoning the differentiability hypothesis (infinite number of geodesics, fractal fluctuations of dimension 2, local irreversibility leading to two-valuedness of derivatives) in terms of such a tool, which takes the form of an operator of "quantum covariant derivation."[170]

This derivation operator, written in terms of complex numbers, completes the transformation of classical mechanics to quantum mechanics, and above all implements scale relativity thanks to an authentic covariance. It enables us to justify definitively the fundamental statement of this approach according to which the motion of quantum "particles" can be described as that of geodesics of a fractal space(time).

What is the covariant form of an equation of geodesics? It is simply the equation of inertial motion, which expresses that the *free particle* moves according to a uniform and rectilinear motion at *constant velocity*, that is, with zero acceleration. *The acceleration is zero*: such is already the form of the equation of geodesics in general relativity. But this simple equation is expressed in terms of the covariant derivative containing the effects of curvature; when those effects are explicated, the equation of Galilean motion is transformed, through the "miracle" of covariance, into an equation generalizing, in Einstein's theory, the equation of Newtonian dynamics in a gravitational field.

A similar result is obtained in scale relativity. An equation keeping the form of that of classical inertial motion (which expresses that the particle moves, locally, in a straight line and at constant velocity), but written using the new "quantum-covariant" (or "scale-covariant") derivative, is transformed, once the expression of it is made explicit, into the Schrödinger equation of a free particle![171] Similar to the effects of gravity appearing as a manifestation of the curved geometry of spacetime, *one thus sees the quantum effects emerge as manifestation of the fractal and nondifferentiable geometry of spacetime.* More generally, *the equation of Newtonian dynamics for a particle in a potential well is transformed into the corresponding Schrödinger equation.*

These results allow us to obtain solutions to the Schrödinger equation by numerical simulation, without having to write the equation explicitly, solely by the description of elementary laws of motion given by scale relativity.[172]

The correspondence principle for momentum and energy (which, in quantum mechanics, associates them with certain differential operators) can also be proven, no longer in the form of a simple correspondence, but indeed with strict equality. Finally, the Born interpretation, which allows one to deduce the probability density of the wave function, is also established in a natural manner. In effect, the set of geodesics can be described as a sort of fluid, of which the density is effectively given by the square of the modulus of the wave function (the imaginary part of the Schrödinger equation is identical to an equation of continuity for this fluid). Over the course of a real experiment, the outcome will be proportional to the density of number of these geodesics, which naturally becomes a density of probability.

Of course, this is only a beginning. The nonrelativistic quantum theory, corresponding to "low" energies (those of atoms) governed by the Schrödinger equation is only a small link in a vast chain which has been developed since the foundation of quantum mechanics in 1925. In progressing toward higher energies (nuclei, elementary particles), one must progress to relativistic situations of velocities on the order of the speed of light.

In this case, it is no longer space alone which is fractal (this hypothesis was sufficient for obtaining the Schrödinger equation), but spacetime. Virtual trajectories thus can move backward not only in space, but in time. This does not pose any causality problem, though, since the sections of the trajectory that move backward in time are simply interpreted, as Feynman has already proposed, as antiparticles. The theory explains virtual pairs of particle-antiparticles, and resolves the problem of the infinite length of trajectories (which could have led to infinite velocities, in contradiction with special relativity): there is indeed divergence of length in space, but also in time, such that the ratio between the two (i.e., the velocity) remains limited by the speed of light c.[173]

Moving forward, the development of the theory should tackle the Klein-Gordon equation (which is the first relativistic equation of quantum mechanics), then the Dirac equation (the relativistic equation for electrons) and it should account for the electromagnetic, weak, and strong fields, and then the electroweak field. Twenty years after the first publication of this book, these successive steps have now been completed in large part, in particular thanks to the work of Marie-Noëlle Célérier.[174] But the vast endeavor which involves translating the different levels of quantum theory in terms of scale relativity and fractal spacetime (and possibly extending it to new realms) has just begun.

Finally, the new theory, as we had initially hoped, allows us to solve the problems of misunderstanding of the quantum theory, not by replacing it, but by refounding it on first principles; and it resolves the problem of quantum/classical incompatibility, but not by establishing a bridge between the two. The proposed solution is obtained with the help of a description which is neither quantum nor classical, but coming from a deeper level, and which can be reduced to classical and quantum descriptions depending on the conditions of motion and scale.

Recall that these conditions are: (i) infinite number of geodesics; (ii) each geodesic is itself fractal; (iii) two-valuedness of derivatives. All three conditions can be derived from nondifferentiable and continuous geometry. One can show that only the three conditions taken together lead to quantum mechanics, while two, one, or none of them lead to various type of classical equations. For example, the case when no condition is fulfilled leads to the usual deterministic Newtonian-like classical theory, while keeping condition (i) leads to a statistical-type classical theory. Therefore, the theory includes a description of the transition from classical laws to quantum laws, by specifying the way the three above conditions may emerge.[175]

Another important consequence of this theory is that the three above conditions, if they are strictly fulfilled in a fully nondifferentiable spacetime (assumed to be exactly implemented at microphysical scales), may also be satisfied approximatively in other, macroscopic situations, for example in chaotic or turbulent phenomena. In this case, one expects a macroscopic, quantum-type "Schrödinger regime" to emerge, which is no longer based on the microscopic Planck constant, but on a macroscopic constant specific to the system. Such a new "macroquantum" physics does not share all the aspects of standard quantum mechanics (as we have recalled, many of them come from elementarity), but nonetheless owns the fundamental aspects of a density of probability which is the square of the modulus of a wave function, itself a solution of a Schrödinger-type equation. In particular, we have suggested, starting in the early nineties,[176] that a fractal medium could play a role similar to a fractal space for the particles that move into it, and that, as a consequence these particle, could acquire some type of macroscopic quantum properties.

Before ending with some examples of applications of the theory to various sciences, let us summarize the steps of its construction, which proceeds mainly by extensions rather than by hypotheses:

- extension of spacetime geometry to a nondifferentiable continuum;
- as a consequence, extension of the description of physical functions to explicitly scale dependent and divergent (i.e., fractal) functions;
- extension to scales of the principle of relativity, by considering resolution as defining the state of scale of the reference system;
- writing the equations of physics in terms of differential equations, not only in position space but also in scale space;
- including the new effects of the fractal and non-differentiable geometry (the three conditions above) in the construction of a new "covariant" derivative;
- writing the equation of motion as an equation of geodesics, which takes the very simple form of the inertia equation in a vacuum in terms of this covariant derivative;
- identifying the wave function with a manifestation of the velocity field of the fluid of fractal geodesics;
- finally, showing that the equation of motion, reexpressed in terms of this wave function, can be integrated under the form of the equations of quantum mechanics.

Part Five
From Elementary Particles to the Large Structures of the Universe

The theory of scale relativity makes it possible to propose solutions to a number of problems in current physics. We will discuss several examples of such propositions, taken from the three domains of application privileged in theory: high-energy physics, cosmology, and the study of self-organized systems in astrophysics. Of course, the novelty of these results necessitates their being considered as preliminary: some of them have been validated, but, as always in science, independent confirmations will be necessary before allowing them as definitively established.

As we shall see, after the first publication of this book twenty years ago, many validations of the first theoretical predictions have been obtained (as for example with exoplanets) and new structures discovered. Moreover, the theory of scale relativity has been applied to several other sciences, such as geosciences, geography, and systems biology.

Chapter 18
Particle Physics

A paradigm shift as profound as that proposed by scale relativity leads to numerous consequences in physics, which must be studied one by one. What is essential is that the theory should be refutable. In the domain of elementary particles, one expects in particular to see growing and non-negligible corrections to standard quantum mechanics at very high energies (beyond the energy of the most massive of elementary fermions, the top quark with mass 174 GeV/c^2), which should be able to be put to the test by the next generation of accelerators.[177]

The most immediate change concerns the relation between scale of mass-energy-momentum and scale of length-time (see figure 24). In today's quantum theory, the two scales are directly inverse to each other, as shown by the Compton or Heisenberg relations. Energy-momentum tends toward infinity when the interval of time-length tends toward zero. In the new laws of special scale relativity, this interval cannot become lower than the Planck scale. However, just as the energy and momentum of a massive particle tend toward infinity when its velocity approaches that of light in special relativity (of motion), the Planck length now corresponds to an infinite energy (in special relativity of scale).

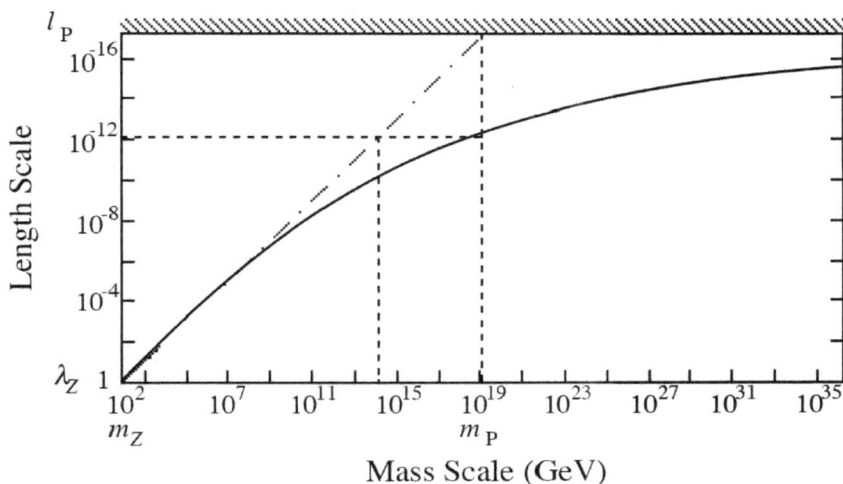

Figure 24 *New relation between scales of length and energy in special scale relativity.* In standard quantum theory, the scales of mass-energy and length-time are inverse to each other (dot-dashed line). In the framework of the new dilation laws (solid curve), the Planck length-scale becomes an unsurpassable and unattainable horizon (shaded area at the top of the graph). It corresponds to an infinite energy. A new scale of length now corresponds to the Planck mass-scale m_p, which is revealed to be, relative to the scale of weak bosons ($\approx$ 100 GeV), the scale of grand unification (λ_p, which is 10^{12} times smaller than the electroweak scale in the minimal standard model).

Grand Unification

In standard theory, the Planck length-scale (10^{-33} cm) corresponds with the Planck energy (10^{19} GeV). This can no longer be the case in special scale relativity. A new universal length-scale, totally constrained by theory (it has no free adjustable parameter), while relative to the scale of observation, is thus naturally introduced. One finds that this scale is 10^{12} times smaller than the length-scale of the gauge bosons of electroweak theory (100 GeV). This new scale corresponds precisely to the scale of grand unification in the minimal standard model. This result means that, in the new framework, the unification of the three fundamental interactions is made at the Planck energy. However, as the

Planck energy is precisely where gravitation becomes on the same order as the other forces, if there should be a unification, it can only be a *total* unification of the four interactions, almost simultaneously (see figure 25).

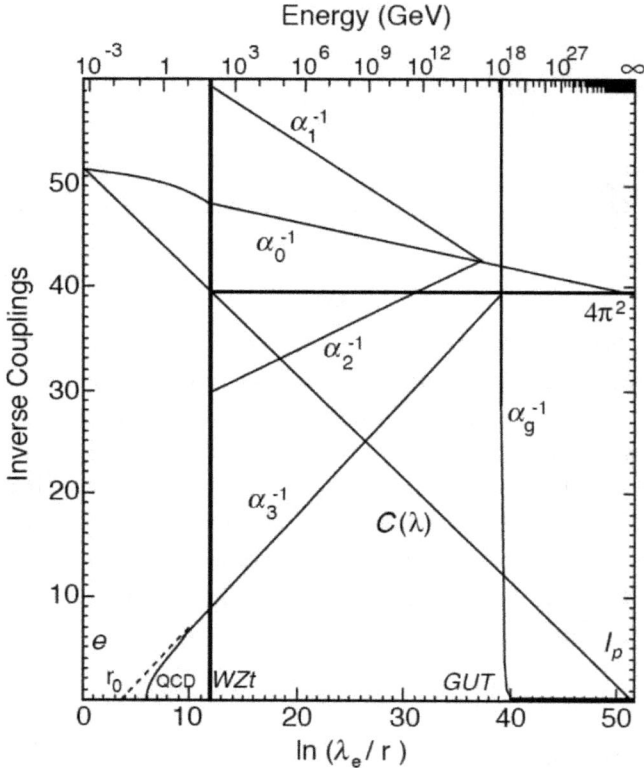

Figure 25 *Variation of charges in the minimal standard model in special scale relativity.* We have shown here the variation in function of the scale of the inverse squares of the charges of fundamental interactions (inverse of running couplings), between the scale of the electron e and that of Planck λ_P (energy scales on top and length-scales below). Two other fundamental scales appear in this diagram, that of grand unification (GUT), which is equated in the new theory to the Planck energy-scale, and that of electroweak unification (WZ). Numerous structures predicted by the new theory are apparent in this figure, for example, the intersections of the running couplings with the line $C(r) = \ln(r/\lambda_P)$ and the special value $4\pi^2$ of the inverse couplings, meaning in particular that the charge value at infinite energy is $1/2\pi$ (I_p in this figure is for the Planck length-scale λ_P).

218

In the current standard model (or its various extensions), one does not obtain such a common unification at the Planck energy scale, but a unification in two steps, first of the three gauge fields (at 10^{14} to 10^{16} GeV), then only later with gravitation at 10^{19} GeV. This result is therefore more satisfying for the mind, although it does not make the problem of explicitly constructing a unified theory any easier.[178] But it would have the advantage of putting an end (at least partially) to one of the fundamental questions of physics: why does the gravitational constant have its value?[179]

Alternatively, one can pose the question by reducing it to the quantum expression of such a force, meaning writing it in natural units given by the product $\hbar c$ of the Planck constant $\hbar$ and the speed of light c. This type of expression allows us to define a dimensionless coupling constant, like for example the fine structure constant in the case of electromagnetic interaction.[180] Applied to the gravitational force, this method introduces the Planck mass as mass unit and implies a coupling constant—that is equal to one![181] The question of the value of G, then, takes a new form: if the Planck mass thus plays the role of a natural unit of mass, should one not expect that this unit be realized in the form of the most elementary and fundamental particles in nature? This is effectively the case. One can consider, in fact, that the most fundamental particles in nature would be the bosons transporting the totally unified interaction, from which would issue at lower energy the particles which are vectors of the disunified fields (photons, weak bosons, and gluons): this unification taking place, in the new framework, directly at the Planck energy, a large part of these bosons would have precisely the Planck mass, and would thus physically realize this universal unit of mass.[182]

Charge Variation in Function of Scale

One of the main results obtained in the application of special scale relativity to particle physics is the suggestion of a solution to one of the most difficult still-open problems in physics: that of the divergence of masses and charges in quantum field theories. Actually, almost all quantities are divergent (infinite) in these theories, but a partial solution has been found, called renormalization. In this method, one replaces the theoretically infinite values of mass, charge and probability densities by their finite observed values, and all the other calculated quantities become finite and agree very precisely with their experimental values. The drawback is that the infinities in masses and charges have not really been solved.

In the new framework of special scale relativity, thanks to the new meaning given to the Planck spacetime scale as a minimal scale, invariant under dilations, the problems of divergences which remain in quantum theory are automatically resolved. The situation is, once again, similar to what happens in special motion relativity. We know that any object of mass m possesses a rest energy mc^2. But in Galilean relativity, the light velocity is assumed to be infinite, so that there is actually a general problem of mass-energy divergence in the classical nonrelativistic theory: this (implicit) problem is solved just by realizing that the velocity of light is finite. The same is true for scales: in the standard theory, the ratio between any length-scale and the null length-scale which corresponds to infinite energy is infinite. In special scale relativity, it is now the finite Planck spacetime scale that corresponds to infinite energy, so that there is no longer any divergence of elementary particle masses and charges.

This enables us to pose in a renewed manner the question of the origin and values of charges (meaning the coupling constants of fundamental interactions). In current theory, these vary explicitly as a function of scale (see figure 19): for example, the fine structure constant (square of the electric charge), which is equal to about 1/137 at the Compton scale of the electron (energy 0.5 MeV), has increased to around 1/129 at the scale of electroweak bosons (energy of around 100 GeV). This variation has raised the possibility of being able to calculate the macroscopic value, observed at low energy, of these "constants" starting from their value at scale zero. Unfortunately, in the standard model, as we have seen, the charges are either zero or infinite at the limit of a zero resolution scale, which has thus far prevented the putting into effect of such a program.

The problem can be resolved in scale relativity, a theory in which naked charges take a finite value, starting from which one can find the observed values of the coupling constants at low energy, and even bring into effect new theoretical predictions for some of them (see below and figure 25).

Nature of the Electric Charge and the Electromagnetic Field

Scale relativity also enables us to clarify the problem of the nature of charge and the interaction fields. In effect, the existence of an "internal" structure of particles (meaning structures inside the fractal geodesics of which they are manifestations) implies symmetries, no longer in spacetime, but in the new dimensions of scale. Similar to how symmetries of space and time constitute the invariant quantities such as energy, linear momentum, and angular momentum, some conserved quantities (here, no matter at what scale) will appear from the fact of these new scale symmetries, quantities which can be equated with charges.

As for fields of interaction—electromagnetism and weak interaction, which at high energies combine to form the electroweak field (coupling constants α_1 and α_2 in figure 25), as well as the strong interaction (α_3 in figure 25)—they are described as "gauge fields" in current quantum theory. The notion of "gauge invariance" plays an essential role in particle physics today, even though some of its aspects remain obscure. It is a case of transformations on wave functions (namely, on their phase), which have no classical counterpart. The importance of this notion holds to the fact that the conservation of charge and even the existence of the electromagnetic field are directly tied to it.

But gauge invariance poses a surprising problem. The nature of the electromagnetic field does not change in a transformation (called "gauge") involving a completely arbitrary function of space and time; nevertheless, this function reappears, coupled to the charge, in the wave function of the electron. In other words, this "arbitrary" function, considered in general as devoid of physical meaning, is the conjugate variable of the charge. This is a very important point, linked to Noether's theorem, which connects the fundamental symmetries of nature with the conservative quantities that these symmetries imply. Hence the uniformity of time leads to the conservation of energy E, which is the conjugate variable of time t. The homogeneity of space leads to the conservation of momentum p (conjugate of position x) and the isotropy of space to the conservation of angular momentum σ (conjugate of angles φ). In the phase of the wave function, one sees the product of these variables and their conjugate.[183]

The gauge function thus holds the key to the nature of charge, similar to how the uniformity of space and time are the keys to the nature of momentum and energy. How could it stay arbitrary under these conditions?

The framework of scale relativity enables us to propose a solution to this problem. The basic idea is to describe the electron as a purely geometric object (a network of fractal geodesics), structured in scale beneath its Compton scale (which is the inverse of its mass). Let us now consider one of these microstructures and observe it (by thought) over the course of a spatial or temporal movement of the electron. It is impossible that this structure remains at the same scale over the course of this movement, since, if that were the case, the scales would no longer be relative, but absolute. From the fact of the relativity of scales, the scale corresponding to this structure ought to vary. In other words, a movement in spacetime must induce an effect of dilation or contraction of the internal scales of the electron.

By mathematically describing such a process, one finds that this dilation field induced by the movements is nothing else than the electromagnetic field, having all its properties. If one now makes the same argument for any other of its internal structures, and one compares the two effects, one finds the relation of gauge invariance. However, in this reconstructed gauge invariance, the function which was arbitrary no longer is, since it is given by a ratio between scales (in logarithm). The gauge invariance can thus be reduced to a scale invariance acting inside the electron geodesics, in the framework of the new meaning given to spatiotemporal resolutions.

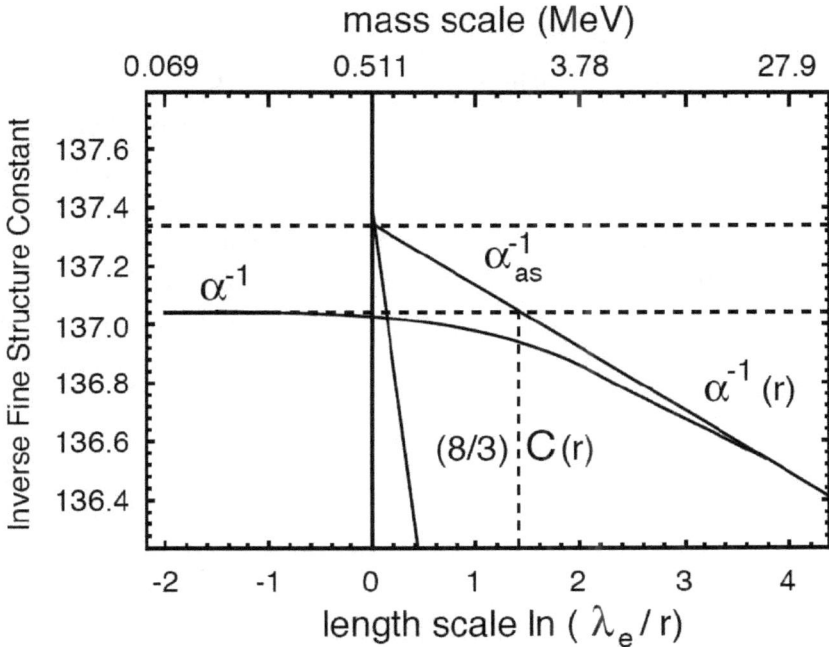

Figure 26 *Relation between mass and charge of the electron.* This diagram is an enlargement of figure 25 around the scale of the electron. The constant α_0 in figure 25 is equal to the fine structure constant α (pictured here) multiplied by 8/3. The observed convergence at the scale of the electron mass between α^{-1} (asymptotic) and 8C/3 where $C(r) = \ln(r/\lambda_P)$, is predicted by the new theory and yields a relation between the mass and the charge of the electron, which enables us to calculate the mass of the electron starting from its charge. As for the charge, it is derived from "running" it from the Planck length-scale (i.e., infinite energy scale), where it is expected from very simple theoretical arguments to have the value $1/2\pi$, which is supported with high precision by experimental values and the knowledge of the charge variation with scale ("running couplings").

This new interpretation of gauge invariance has no consequence in Galilean scale relativity (in which resolutions have no lower bound).

On the other hand, it enables us to obtain new results in the framework of Lorentzian scale laws where the Planck length-scale becomes a minimal scale. Let us recall, in fact, that the conditions of quantization in quantum mechanics come from the existence of limit conditions. For example, it is the constraint that angles always vary between 0 and 360° which necessitates the universal quantization of angular momenta. It is the same here. The limitation by the Planck scale of relations between possible scale ratios implies the quantization of charges and enables us to obtain new relations between masses and charges, which are well verified experimentally (see the structures of figures 25 and 26). Through this process, both the mass of the electron (511 keV, from its ratio with the Planck mass) and its charge squared (1/137.04 in dimensionless units) can be theoretically predicted with high precision.

Chapter 19
Cosmology

The applications of scale relativity to cosmology occur in two domains.[184] One studies the consequences of the new dilation laws on the microscopic scale for the description of the primordial universe (big bang theory). The other considers that the Lorentzian scale laws can also be applied to very large scales, which would lead to the introduction of a maximal scale of resolution, unsurpassable and invariant under all dilations, possessing the physical properties of infinity (the reverse of the Planck spacetime scale, which possesses the properties of zero).

The primordial universe is one of the natural domains of applying special scale relativity, for multiple convergent reasons: because of the great temperatures and high energies of the first instants, big bang theory has already connected quantum theory and cosmology; the expansion of the universe can itself be described as a universal scale law (the dilation over the course of time of all the inter-distances between galaxies); moreover, scale relativity brings important new corrections to very small scales, spatial as well as temporal, which is precisely the domain of the first instants of the Universe.

The Problem of the Origin

One first consequence concerns the problem of the origin. The point zero, spatial or temporal, no longer having any physical meaning (no longer existing!) in scale relativity, we can no longer start expansion at the moment $t = 0$. The singularity poses a problem for certain cosmological models: for example, in the open hyperbolic model (the one which seems favored by most observations today), the universe is infinite in its simplest topology, and this is true at every instant t greater than zero, no matter how small, while its spatial part disappears "all at once" at the moment $t = 0$. In scale relativity, this is no longer a problem, since the expansion would start asymptotically from the Planck length and time-scale, for which spacetime is completely degenerate (Planck time is a "horizon").

But, in any case, it seems premature to try to understand this period, knowing that we are doubtless very far from understanding its laws. It is perhaps insufficient to construct a theory of quantum gravity to comprehend this domain. In fact, such an attempt (which has not finished to this day) rests on the idea that gravity should be described by quantum field theory when quantum effects and gravitational effects become of the same order. However, if there is no doubt that the current theory of gravity becomes insufficient at the scale of Planck energy, one often forgets that it is probably also true for quantum theory itself! All of our physical theories must be "reset to zero" at this energy, and we are probably still very far from even catching a glimpse of what this unified theory would be.

One must add to this that, during the earliest moments of the big bang, there is no longer any static scale to which to relate the intervals of length and time: only the relations of dilation and contraction can be defined. In other words, to speak of periods of time such as 10^{-30} seconds after the singularity or a radius of the universe of 10^{-25} cm is probably meaningless.

Horizon and Causality

One of the problems for which scale relativity proposes a novel solution is that of horizon and causality. Recall its statement: the universe is filled with a bath of photons at the temperature of 2.73 Kelvin, a fossil of the earliest moments. If one observes this cosmological blackbody radiation in two opposing directions into space, one finds that its temperature (once correcting for our own movement relative to it) is practically the same, with differences around only 10^{-5} or so. Yet due to the existence of the initial singularity, the regions of the universe from which this radiation arises have never been able to be causally connected in their own past. In these conditions, how can they show the same temperature (knowing that a blackbody has radiation in thermodynamic equilibrium with its sources, and that its thermalization is made precisely by continual interactions)? The response given in general to this problem is that the primordial universe could have undergone a phase of inflation (meaning exponential expansion) which could have allowed us to connect all the regions observed today. But this solution is both ad hoc and partial. It only establishes a connection among one another for regions of the universe observable today, and would no longer be valid in a very distant future. Moreover, it necessitates the introduction of a new field of interaction (and even a series of fields), completely unobserved and for which nothing otherwise seems to confirm its existence.

In scale relativity, the problem is no longer present, due to the new laws of dilation which take the form of the Lorentz transformation (see figure 27). These laws require that any distance, measured at the Planck resolution, would have the Planck length. Such a statement can seem, at first glance, extremely paradoxical, if not totally absurd. The size of this book, the diameter of the Earth, our distance to the Andromeda nebula, the universe itself would be reduced (even today, not simply during the big bang) to the Planck scale! But a simple thought experiment shows that it is indeed the case, without any contradiction. To make an *effective* measurement of any one of these lengths at the Planck resolution, it would be necessary to construct a ruler for which the gradations are separated by the Planck length. But the generalized Heisenberg relations in the new framework state that it would take an infinite energy-momentum just to mark these gradations, or to compare them to the lengths to be measured. Where can we find such energy? Only the universe in its entirety can provide it (and again, only if it is infinite). To make such a measurement, one would have to inject the entire universe into the Planck scale, which returns us finally to the initial statement (which means, of course, that ultimately it is impossible to make explicit measurements at the Planck resolution).

Applied to the first moments, this new structure implies that all the points in the universe are automatically connected at the Planck era, which solves the problem of causality (figure 27).

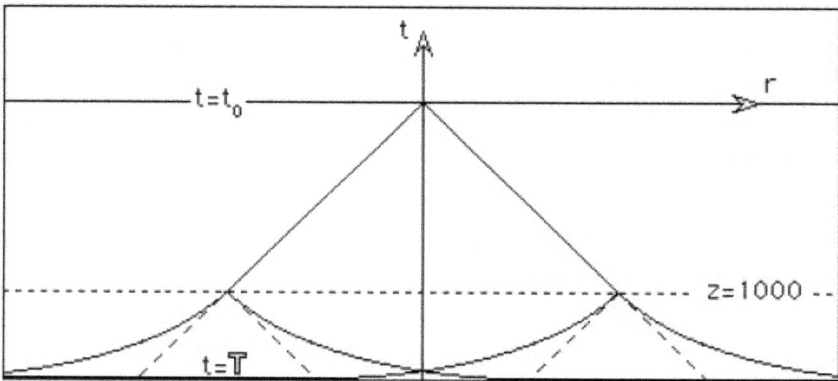

Figure 27 *Widening of light cones approaching the Planck era in special scale relativity.* In the new framework, all the points of the universe could have communicated with each other in the past. The horizontal axis corresponds to space and the vertical to time. The base of the diagram is the big bang (which starts asymptotically in the new theory starting from the Planck time). The dotted line (z = 1,000) corresponds to the period of the combination of nuclei and electrons into atoms. The current epoch is at the top (t= t_0).

Maximal Scale and Cosmological Constant

One of the principal results of the relativistic approach concerns the universality of the Lorentz transformation. The application of the principle of relativity to the dilation laws at small scale leads to the introduction of a minimal scale, invariant under dilations. But what if one applied the same principles to very large scales, referring to cosmology? One then obtains a similar result: the appearance of a maximal scale, invariant under dilations, a horizon for the possible resolutions in nature, which would possess the physical properties of infinity in the same way that the minimal scale possesses the properties of zero. Here again, one must understand that it is in no way a barrier, a limit, or a finitude in the ordinary sense of the term: to the question of if we can consider a given scale, and multiply it by 2, then by 2, and so forth to infinity, the answer still remains yes. On the other hand, it is the result of the successive multiplications which is now found to be limited. The two first products will yield a result a little less than 4, the third less than 8, and so on. It is a limit for possible resolutions in nature, which only have meaning in the new paradigm where those resolutions become explicit variables, intrinsic to the nature of spacetime, which again becomes fractal at very large scales.

If it is natural to identify the minimal scale as the scale of the Planck length λ_P, what would it be for the maximal scale? Does there already exist a scale of length, produced by theory or observation, of which the properties could be reinterpreted (and better understood) in terms of this maximal scale?

Einstein's general relativity in fact provides an invariant scale whose meaning is not clear and over which much ink has been spilt: this is the scale of the cosmological constant. This constant Λ was introduced by Einstein in 1917 in his equations for the gravitational field, to allow for static cosmological solutions (in agreement with the results of stellar observations at the time, which were limited to our own galaxy, which is globally static) satisfying Mach's principle. The identification of spiral nebulae as extragalactic objects in 1924, and then the discovery of the expansion of the universe seemed to make this constant no longer necessary. Nevertheless, Élie Cartan had meanwhile shown that the equations of Einstein with a cosmological constant constituted the general solution to the problem which Einstein had posed, and that there was not any *a priori* reason to exclude it.

The cosmological constant Λ in Einstein's equations has the dimension of a curvature (which is one of the tools allowing us to describe curved geometry: for example, the curvature of a sphere is positive, it is the inverse of the square of its radius, while the curvature of a hyperbolic space is negative). Therefore it is the inverse of the square of a certain cosmic length: $\Lambda = 1/L^2$. To suppose that the cosmological constant is vanishing means to make the cosmic scale L *a priori* infinite. The cosmological constant Λ is apparently very small: the observational limits at the time of the first publication of this book indicated that it is lower than 3×10^{-52} m^{-2}; it is now well measured thanks to the WMAP and *Planck* satellites, and its value is around 1.1×10^{-52} m^{-2}. This very small number simply manifests the fact that the corresponding cosmic length is very large, namely it characterizes cosmological scales (see figure 28).

Several arguments enable us to equate the new scale of maximal resolution with this cosmic length. The first is that it is strange to see an invariant length, static like L is, defined at a scale of length on the order of the size of the universe, where all real physical objects are *a priori* drawn to by expansion. The second is a calculation of quantum gravity made by Stephen Hawking, according to which the probability of possible values of Λ is maximal for the value zero, meaning L is infinite. But this calculation is performed in a framework where the laws of dilation are the ordinary laws (which correspond to a "Galilean" scale relativity). Generalized to Lorentzian laws, one expects that the most probable value of L is now given by the scale that plays the role of infinity in the new framework, meaning the maximal scale, invariant under dilations.

The existence of a minimal scale (the Planck scale λ_p) and a maximal scale (the cosmic length L), finally introduces into theory a pure number K, invariant, given by the ratio of these two lengths $K =$

L/λ_p, for which the value is now (2018) precisely known as $K = 5.9$ x 10^{60}. The existence of this number, which corresponds to a ratio of maximal possible dilation between resolutions, allows the understanding of certain empirical cosmological relations (called the large number coincidence of Eddington-Dirac), up until now obscure. Let us see how.

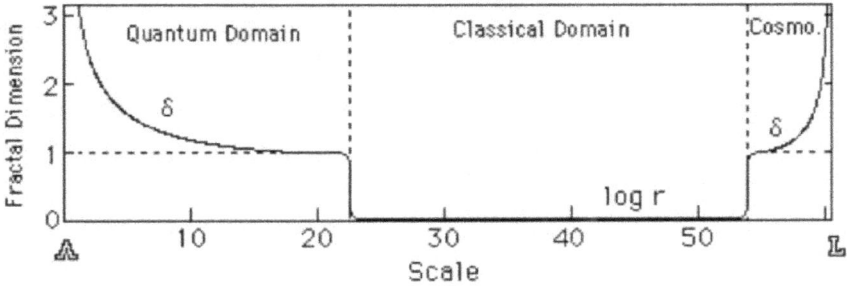

Figure 28 *Schematic variation, as a function of the resolution, of the effective fractal dimension (given by D = 1 + δ) of the geodesics of spacetime in special scale relativity.* There appears in this framework a scale of minimal resolution, identical to the Planck scale, and a maximal scale, tied to the cosmological constant. The scale symmetry is broken in two (non-absolute) transition scales which divide the scale space into three domains (at least, since this is just a schematic and simplified view): a classical, intermediate, domain, where spacetime does not explicitly depend on resolutions since the laws of motion are dominant with respect to the scale laws (but in this domain, many systems may be fractal on a large range of scales); and two asymptotic domains at very small or very great scales (microscopic and cosmological) where the scale laws dominate the laws of motion, which makes explicit the underlying fractal structure of spacetime.

Energy Density of the Quantum Vacuum

Yakov Zel'dovich tried to elucidate the nature of the cosmological constant starting from the following observation (initially made by Georges Lemaître). In the Einstein equations, the cosmological constant *a priori* plays a geometric role. But it turns out that its contribution has the same form as that given by the energy of the quantum vacuum. Recall that the vacuum, in quantum mechanics, is not nothingness: It is the minimum state of energy of a field, which is *a priori* non-null (this is what allows the electron not to crash into the nucleus in an atom). From this came the idea that the cosmological constant was the manifestation of the energy density of all the zero points (the vacua) of the different fields that exist in nature. More generally, the true cosmological constant should be the sum of an eventual geometric constant and of the density of the vacuum. Unfortunately, if one wishes to calculate this energy density, one finds it to be infinite! Perhaps one should only integrate the zero points of fields just up to the Planck scale, which represents a sort of barrier in the standard model of physics: one then finds an effective cosmological constant 10^{120} times larger than its astrophysical estimations!

The proposed solution in scale relativity (see figure 29) consists of describing the vacuum as a fractal (i.e., as scale dependent and diverging toward small scales) and of considering its energy density not as a number, but as an explicit function of the scale, in agreement with the basic method of this theory. This explicit scale dependence of the vacuum energy density is also just what one expects from quantum mechanics, through the Heisenberg relations. There is then no reason to attribute to the universe (at the cosmological scale) the value of the energy density calculated at the Planck scale. In fact, the energy is defined as a close additive constant, so that the quantum vacuum can very well be reduced to a zero energy (it is thus said to be "renormalized"). On the other hand, in terms of what concerns the *fluctuations* of the vacuum energy, if those are cancelled out on average, their average square cannot be cancelled out due to the fact of the Heisenberg relations. Yet these fluctuations will gravitationally interact with each other, and will produce a self-energy of gravitational coupling which itself cannot be avoided. The density of this self-energy varies as a function of the scale as inverse of the square of volume, which enables us to propose an explanation of the Dirac large numbers relation. This explanation states that the scale of elementary particles is found as a third of the scale of the universe in the scale space in logarithmic notation (see figures 3 and 29), which one effectively obtains by the calculations of scale relativity. One then obtains an acceptable value for the cosmological constant (1.36×10^{-52} m^{-2}), thus the scale of maximal length.[185] I concluded the first version (1998, in French) of this book with the words: "We will have to await the direct measurements of astrophysics to confirm or deny this possibility."

Twenty years later (2018), the value of the cosmological constant (nowadays sometimes called "dark energy") is now precisely known from astronomical observations (a Nobel Prize has been given to this result in 2011), and the theoretical argument has been improved. The agreement between the theoretically expected value (reduced cosmological constant 0.3115 ± 0.0001) and the observational value (0.318 ± 0.012 from the *Planck* mission) is very good.

Let us summarize this more complete argument. It relies on applying the expansion of the universe, not only to real matter, but also to the virtual particles which constitute the quantum vacuum. In Dirac's description of the vacuum, it is like a "sea," totally filled with particles of negative energy. If a particle from the vacuum jumps to positive energy because of quantum fluctuations, it leaves a hole in the Dirac sea, and the couple (particle, hole) will be seen as a particle-antiparticle pair. The interdistance between the Dirac sea particles, and therefore between the pair members, is expected to increase with time because of the expansion of the universe. As we have previously seen, the self-energy density of the pairs can always be renormalized, but not the gravitational self-energy density (thanks to Heisenberg's relations), which is therefore the best candidate for being the source of the "dark energy." But this gravitational self-energy density of pairs (i.e., of energy fluctuations) decreases very quickly (as $1/r^6$) when the interdistance r between the pair members increases. As a consequence, it is not able to constitute today an effective cosmological constant.

Indeed, two of the main properties of the vacuum allowing it to appear as an (invariant) cosmological constant is first to be Lorentz invariant (which it is), but also an adiabatic invariant: namely, it should not change under a dilation or a contraction. But the expansion of the universe is just a dilation of the universe over time. Therefore, the vacuum energy densities of almost all components of the Dirac sea are unable to make a cosmological constant.

However, there is one class of elementary particles for which the situation is different: quarks. Quarks are said to be confined in the nucleons and in other composed particles like the pion. This means that one has never seen a free quark, and that this is probably impossible. Why? Assume you want to separate the two quarks making a pion by pulling them apart. In order to do that, you must use energy. The force that links the two quarks together is like a string, and you must use enough energy to break the string. But one finds that this string-breaking energy is larger than the energy needed to create a new quark-antiquark pair. Thus, while you wanted to get two separated free quarks, you actually end with two separated pairs of quarks, while the interdistance between the quarks in each pair remains unchanged!

Applied to the Dirac sea of quarks, this argument means that the interdistance between quarks of the vacuum takes its maximal value. No composed particles contribute to the vacuum; it is made of truly elementary particles, in this case quarks. When the expansion of the universe pulls them apart, the confinement energy leads to particle creation, which keeps the density of particles constant. In other words, the quark vacuum is frozen.

Having arrived at this point, there are still two problems to be solved: (i) the gravitational self-energy density of quantum fluctuations is negative, while dark energy is positive; (ii) this gravitational energy is of gravitational origin, and it has therefore no reason to be itself a source of gravitation (while the cosmological constant enters into Einstein's equations of the gravitational field). Both problems are solved by an implementation, introduced by Sciama, of what Einstein has called "Mach's principle," which is the "postulate of the relativity of all inertia." This means that any body, ultimately, should be free (this is also the conclusion derived, by other means, in the theory of scale relativity). Concretely, this principle implies that the total energy of any body or entity should be null. When it is applied in the rest frame of a massive body, it means that the sum of the rest mass mc^2 of this body and of its gravitational coupling with all the bodies of the universe should vanish. Due to the equivalence principle, the mass of the body disappears from this equation, which becomes a cosmological relation between the "mass" and the "radius" of the universe.[186] But the main point here is that, in this Machian relation, the gravitational potential energy is found to be just the opposite of the body's mass, which is the source of gravitation.

Mach's principle should apply to the quantum vacuum as well. If not, the vacuum energy density (created from the gravitational potential energy of fluctuations) would contradict the freeness of any entity in physics. Therefore one expects it to be cancelled by a "dark energy," which is just its opposite (thus with positive sign as expected) and which may be a source of gravitation, as mass is.

The last step toward a solution to the cosmological constant problem amounts to having an estimate of the interdistance between the Dirac sea quarks at which this freezing occurs. Concerning real particles, the lowest energy particles composed of quarks are pions (remember that lower energy-momentum corresponds to larger time-length in quantum mechanics). Charged pions have a small part of their mass coming from electromagnetic energy, while neutral pions can be expected to yield an optimal knowledge of pure confinement energy (which is probably of quantum chromodynamical [QCD] nature). Therefore the effective mass of quarks in the neutral pion (67.49 MeV) and its associated length-scale (r_0 = 2.92 fm, see Fig.29) provides a theoretically founded numerical estimate of the dark energy/cosmological constant (reduced constant 0.3115±0.0001).

As we have recalled, it is in very good agreement with the observed value (0.318±0.012). Moreover, being more precise by a factor of $\approx$ 100, this prediction could be checked again in the future (accounting also for the fact that other contributions may exist at this level of precision). Note also that it is not fully theoretical, since it relies on the experimentally observed value of the pion mass. However, being a composite particle, it is possible to derive its mass from the quark masses (through Goldstone boson models, similar to the Higgs mechanism, but in QCD instead of electroweak theory). The problem is that the quark masses are unknown, so that these models are used in the reverse way, by deriving the quark masses from the composed particle masses. If one day a theoretical prediction can be made for the quark masses, the prediction of the cosmological constant value by this process would become fully theoretical.

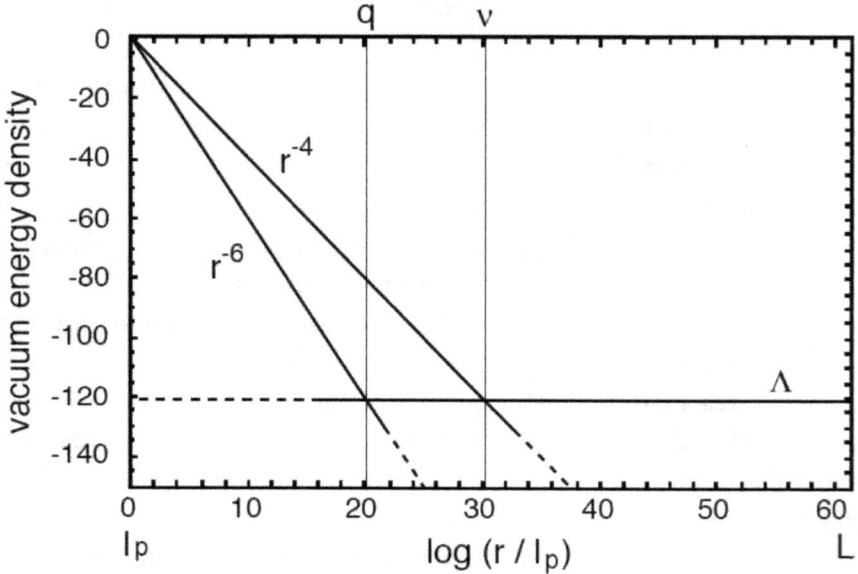

Figure 29 *Variation as a function of scale of the effective cosmological constant (proportional to the energy density of the quantum vacuum).* It is calculated as the sum of a geometric constant, ρ_{cosm}, and of the gravitational self-energy of the fluctuations of the quantum vacuum, which varies as $1/r^6$. These two terms meet in the domain of the scale of elementary particles ($\approx$ 1 MeV to 100 GeV). The fact that the energy density varies as the inverse of the sixth power of the scale, while the cosmological constant varies as the inverse of the square of the cosmic length, allows us to explain the empirical relation of Dirac $(L/\lambda_P) = (r_0/\lambda_P)^3$, where λ_P denotes the Planck length and r_0 the elementary particle scale where the quantum vacuum is "frozen." It can be most probably identified with the scale of the effective mass of quarks in the neutral pion, corresponding to an energy of 67.5 MeV.

Chapter 20
Formation and Evolution of Structures

To finish, we arrive now at one of the propositions of the new approach for which the consequences could be the greatest: the idea that the formalism of scale relativity (including some of its quantum-type method) can be applied (over relatively large scales of time and at the cost of a different interpretation) to systems that had been considered until now to be strictly classical.

The reason for such a proposition is the following: as I have briefly shown in Part Four, the Schrödinger equation can be obtained in a very general manner starting from the fundamental equation of dynamics. In all the situations where one can no longer attribute well-defined individual trajectories to bodies, since they can follow a large number of potential trajectories; where each one of these virtual trajectories is a fractal curve of dimension 2 (which corresponds to a total loss of information); where, finally, there is an irreversibility at small scales of time; in all these cases, one can replace classical deterministic dynamics with a statistical description in terms of probability amplitudes (another name for wave functions) and the Schrödinger equation. The square of this amplitude will then give the probability density of the structure under consideration (that the individual particles will only cross). Garnet Ord (who also introduced the concept of fractal spacetime at the beginning of the 1980s for understanding quantum mechanics) has also arrived at a similar conclusion concerning the possible universality of the Schrödinger equation.

I must of course insist again on the fact that the corresponding theory is *not* standard quantum mechanics, such as it is applied to atoms and particles. In the microphysical domain where it describes elementary objects, it is associated with an interpretation and a theory of measurement which have no reason to remain valid in macrophysics. In particular, the "macroquantum" application of the scale relativity theory is not expected, in general, to involve extreme quantum behaviors such as the indiscernibility of identical particles or quantum entanglement.

The advantage of the new description is that it can be done in terms of wave functions, solutions of a generalized Schrödinger equation, from which one can deduce a probability. This description is then interpreted as the tendency for a system to form structures. By abandoning the hope of following individual trajectories, we gain the possibility of describing a self-organization, a morphogenesis of complex systems. In fact, one of the properties of quantum theory is to predict the characteristic structures of well-determined morphology (for example, atomic orbitals), in a manner that is tied to boundary conditions, and to the various spatiotemporal constraints to which the system is subjected (external forces, symmetries). Stationary solutions can be obtained independently from any initial condition, which enables us to solve problems of the "chicken and egg" variety.

Gravitational Structures

It seemed logical to try, initially, to validate the new theory in the area where chaos was for the first time demonstrated, celestial mechanics. This established all the more so that our own solar system seems to show numerous structures, all of which are not satisfactorily explained by the standard model of formation (distribution of the distances between planets, their angular momentum, their mass, etc.).

It is a matter of applying the new theory to bodies immersed in a gravitational field. This imposes new constraints, which will render yet more distinct the theory of its microphysical counterpart. The primary constraint is the equivalence principle of Einstein, which should remain valid in the new framework. This is translated by the fact that, in the equation of generalized geodesics (which is transformed into a Schrödinger equation), the inertial mass of the object whose motion one is following should disappear. Moreover, in the case of the Kepler problem (that of a body orbiting a central body), one expects that the fractal fluctuations increase with the active gravitational mass of the central body, as is already the case for the curvature of spacetime. As a result of these two constraints, *the constant w which appears in the Schrödinger equation is expected to have the dimension of a velocity*, instead of the Planck constant of action, $\hbar$, having the dimension of an angular momentum, which appears in standard quantum mechanics applied to microphysics. One thus foresees a universal quantization of the *velocities* of astronomical bodies: as we shall see, this prediction is now verified by observational data from our own solar system up to cosmological scales.

The Kepler Problem

In applying the new theory to Kepler's problem of a body orbiting a central potential (which can correspond to the planets around a star or again to binary star systems, to galaxies, or to other celestial bodies), one finds that, contrary to the "predictions" of classical theory, all the distances are no longer equiprobable. More precisely, the classical theory makes no prediction in that case, which we translate by equal probability. It is able to predict the position of a planet starting from known initial conditions of position and velocity, but it has no answer to the general question: "What is the distance of a planet to its star?" This is the opposite of the quantum description of an electron in an atom: the theory can give the probability of presence of the electron at a certain distance of the nucleus, independently of any initial condition (but dependent on its quantized energy). The distribution of probabilities of position of planets shows marked peaks for quantized values of the distance. As a function of a certain quantum integer number n, one predicts that the most probable distances should vary as n^2 and the velocities as $v_n = w/n$.

The Solar System

The first Keplerian system on which to put such a prediction to the test is clearly our own solar system. The intuition of the existence of a law of structuring the distances between planets goes back to Kepler. Numerous empirical propositions have been made (Titius-Bode laws), but have not been very convincing or significant, since there are too few constraints in this kind of law (too many parameters for too few objects).

Our prediction of an n^2 law sheds new light upon this question, since it is made with only one free parameter, the constant w_0. Moreover, as we shall see, the value of this constant, $w_0 \approx 150$ km/s, while not predicted theoretically,[187] can be derived from extragalactic observations, which are completely independent of planetary data and, however, yield the same structures, with the same numerical values for velocities (although the distances differ by a factor 10^{12}!). It has already been noted elsewhere that empirical laws varying as n^2 provide a much better alignment with our planetary systems than the scale laws (Pecker and Schatzman). One can also recall in this regard the attempt of Herbert Jehle, who had proposed as early as 1938 to apply the recent development of quantum mechanics to the solar system (but without any justification), and that of Philippe Blanchard, of the University of Bielefeld in Germany, who used Edward Nelson's stochastic mechanics as description of the diffusion process of the initial disc (but in a different manner than the proposition made here).

I have applied the new scale relativity approach to the solar system in collaboration with Gérard Schumacher, Jean Gay, Patrick Galopeau, and Eric Lefèvre.[188] It is in fact sufficient to calculate the velocities of the planets of the inner solar system (the telluric planets) to verify that our theoretical prediction is accurate in its smallest details (see figure 30). The velocities of the inner solar system's planets are effectively given by $v_n = w_0/n$ km/s, where $w_0 = 144.3 \pm 1.2$ km/s, and where Mercury, Venus, the Earth, and Mars respectively take the positions $n = 3, 4, 5$, and 6: ≈ 48 km/s for Mercury, ≈ 36 km/s for Venus, ≈ 29 for the Earth, and ≈ 24 for Mars.

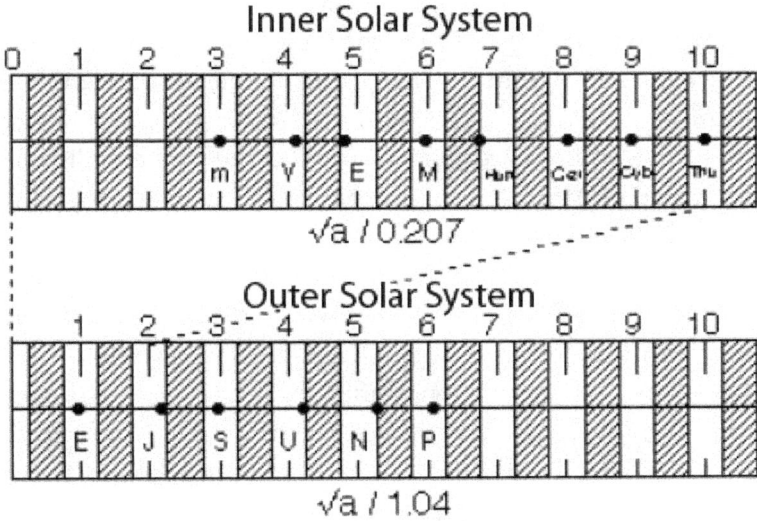

Figure 30 *Fit between theory and observations in our solar system.* the figure compares the observed positions of planets with theoretical prediction (white: peaks of probability density; grey: minima of probability density). The internal solar system in its entirety is the orbital $n = 1$ in the outer solar system, which allows us to reconcile the constants of the two systems.

This law extends up to the asteroid belt, where the principal peaks (in mass) correspond to $n = 7$ to 10. Its validity for the system of satellites around the giant planets has been recently confirmed with a high statistical significance.[189]

But the most remarkable result is that, according to this law, Mercury is not the first potential planet in the solar system, but the third. In particular, the fundamental level ($n=1$) predicted by the new theory is expected to lie at 0.043 Astronomical Unit by solar mass (AU, the distance between the Earth and the Sun). Moreover, this theory is not specific to our own solar system, but should apply to any planetary system formed under similar conditions. As we shall see hereafter, this theory has been able to anticipate the discovery of extrasolar planetary systems and to predict (before their discovery starting in 1995) the existence of fundamental structures for these systems, the first of which is a peak of probability density lying just on this fundamental orbital.

In our solar system, two intra-mercurial orbitals are thus predicted, one at ≈ 0.17 AU ($n = 2$) and the other at ≈ 0.043 AU from the Sun ($n = 1$). The existence of a small planet or of an asteroid belt at 0.17 AU is not out of the question.[190] The mass of a planet should be lower than 1/1,000 of that of the Earth, to not perturb the well-established result of the advance of Mercury's perihelion predicted by Einstein's general relativity (42.98 arcseconds per century, the observed value being 43.11 ± 0.21). No small planet has been found, but there is indirect evidence for a possible asteroid belt at this distance from the Sun, since the extrapolation of the orbits of many Earth-crossing asteroids has shown that they come from this region. Similarly, if it seems very improbable that a small body could have survived at 0.04 AU from the Sun (it would have quickly evaporated), there have been repeated indications of the existence of transient dust density peaks just at this distance from the Sun.[191]

Many other structures in the solar system can be accounted for by the "macroquantum" approach. Some of them were known but misunderstood, but also completely new structures have been predicted, then subsequently validated by new observations.

The mass distribution of planets in the inner and outer solar system (decreasing toward the Sun and toward the exterior) can be accounted for by the shapes of the solutions ("orbitals") to the equation of dynamics, which has taken a Schrödinger-type form (due to the fractality of the chaotic orbits). These shapes provide the density distribution of the initial protoplanetary disk, which is subsequently fragmented into substructures finally giving rise to the planets with their observed mass distribution: decreasing toward the interior and the exterior, implying the existence of asteroid belts instead of accreted planets to the exterior, namely, the Mars-Jupiter belt for the inner solar system and the Kuiper belt for the outer one.

The new fractal approach can be applied to the Sun itself. The Keplerian velocity at its radius (w_{sun} = 437 km/s) is very close to three times the velocity of the fundamental orbital. Moreover, the concept of a macroquantum theory naturally leads to introducing the equivalent of the de Broglie wavelength and period (but with a macroscopic constant which is no longer the Planck constant, but is now linked to the Sun mass and to the above velocity w_{sun}). The remarkable result is that this macroscopic de Broglie period yields exactly the Sun cycle period of 11 years, which was up to now unexplained. The same formula also accounts very well for the recently observed magnetic and spot cycle of other nearby solar-like stars.[192]

Structuring has been predicted and validated for many other objects in the solar system, such as perihelia of sungrazers (comets which approach very close to the Sun and often fall on it); the Mars-Jupiter belt; obliquities and inclinations of planets and satellites; and even space debris around the Earth.

But one of the most impressive successes of the theory concerns the Kuiper belt (the scattered and very distant one), since, as for exoplanets, the prediction of structures was made at the beginning of the 1990s, well before the objects themselves were discovered. The first prediction amounts to just extrapolating the outer solar system sequence in n^2, as can be seen in Fig. 30 (lower part). The inner solar system is n = 1, Jupiter, Saturn, Uranus and Neptune are n = 2 to 5, and Pluto n = 6. We now know that Pluto is not alone, and that the main Kuiper belt shows a large density peak just at this distance of 40 AU. But we can see on this figure (dating from the beginning of the nineties and given in the first version of this book, then still empty) that other zones of high probability are expected beyond this distance (n = 7 to 10).

As can be seen in Fig. 31, the scattered Kuiper belt objects (SKBOs) discovered since this book's original publication support this prediction in a remarkable way. Moreover, it is noticeable that the dwarf planet Eris, the discovery of which is the cause for Pluto, less massive than Eris, being demoted to the status of dwarf planet, has a semimajor axis that lies just in the density peak $n = 8$, at a distance of 68 AU.

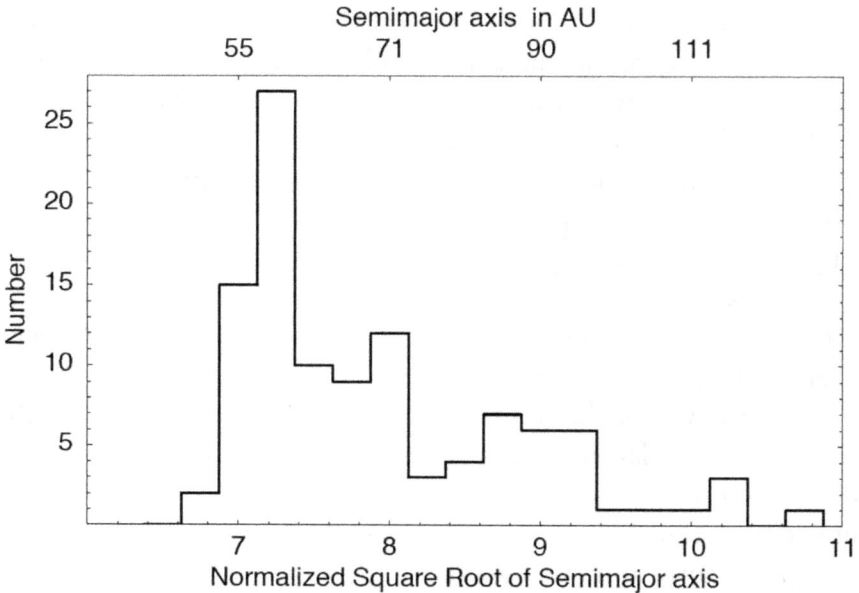

Figure 31 *Distribution of the semimajor axis of Scattered Kuiper Belt Objects, compared with the theoretical expectation of probability density peaks for the outer solar system.* (The very high Kuiper belt peak at $n=6$ is not shown). One theoretically expects probability peaks for integer values of the variable $\sqrt{(a/1.115)}$, a being the semimajor axis of the orbit in AU. The probability of obtaining such an agreement between prediction and observation by chance is only 2×10^{-4}.

This is not all. A new hierarchical level of the outer solar system can be expected for the SKBO population, by taking their main peak at 57 AU as the fundamental level. One therefore expects new probability peaks for semimajor axes $57\, n^2$ AU, i.e. 228 AU ($n = 2$), 513 AU ($n = 3$), 912 AU ($n = 4$), 1425 AU ($n = 5$), etc. This new level of hierarchy has indeed been discovered these last years with the dwarf planet Sedna, having a semimajor axis of 509 AU, which agrees closely with the $n=3$ density peak. Moreover, several objects have been found around 220 AU ($n=2$) and one around 910 AU ($n=4$).

The solar system is therefore structured in a self-similar way over at least five levels of hierarchy (Sun, intramercurial zone, inner solar system, outer solar system, very distant SKBOs), with each of these levels themselves showing well-defined structuring in close agreement with the new "Schrödinger" approach (based on the chaotic dynamics of these subsystems on long time scales).

Extrasolar Planets

The scale relativity approach has been successfully applied to the study of extrasolar planetary systems.[193] At the time of this theoretical prediction (beginning of the nineties[194]), extrasolar planetary systems had not yet been discovered. But based on the universality of the proposed theoretical approach, we were able to conclude that these planetary systems, once discovered, should show universal structuring: in particular, we expected exoplanets to lie preferentially in the fundamental 'orbital' at an orbital velocity of about 150 km/s (corresponding to a semimajor axis of 0.043 UA/solar mass). This prediction has been verified in an extraordinary way, since the first exoplanet discovered around a solar-like star, 51 Peg, has been found to lie precisely at this expected fundamental distance, followed by many others. Today (2018), more than 3,000 exoplanets are now known, and almost one third of these planets have been found in the predicted peak of probability density.

I show in Fig. 32 (established before 1998 and given in the first edition of the present book) the observed distances to their star of the first discovered exoplanets, compared with the solar system planets and with the zones of expected high probability. Then in Fig. 33, the observed distribution in 2008: the n^2 law was still supported with a very low probability to be obtained by chance of only 4×10^{-7} and the fundamental level peak has become clearly dominant. There is a clear peak for $n = 7$, while this density peak is suppressed in our own solar system due to resonances with Jupiter. This is different from the case of the $n=8$ and $n=9$ peaks, which are manifested in the solar system as the main masses of dwarf planets (Ceres and Hygeia) in the Mars-Jupiter asteroid belt.

Today (2018), there are a large number of different planetary systems discovered by various methods, which have a tendency to smooth out the subtle structures. However, as can be shown in Fig. 34, the main density peak predicted from the solar system data to lie around 140-150 km/s is still present in a very clear way, thus offering a strong validation of the scale relativity theory (in its macroquantum version applied to the formation and evolution of planetary systems). The paradox is that such a scale theory, which places the scaling effect not at the level of phenomena, but at the fundamental level of spacetime itself, leads to a crystal-like regular quantization inside each subsystem (not of the positions themselves, but of their peak of probability density). However, a self-similar scaling is recovered between the subsystems, as we have seen for our own solar system.

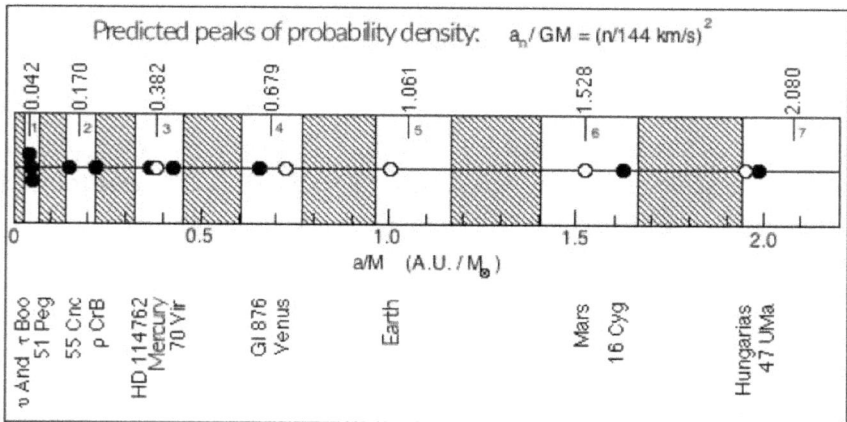

Figure 32 *Fit between theoretical predictions and observations for the first discovered exoplanets (1996).* This diagram (given in the 1998 edition of this book) compares, for the inner solar system and the first exoplanets to have been found, the predicted values at the observed values of the semimajor axes of planetary orbits (the zones of high probability are in white, those of low probability are shaded). The peaks of probability are not adjusted, but are *a priori* fixed (they correspond to the constant $w_0 = 144$ km/s). The probability of obtaining such a configuration by chance is less than 1/10,000.

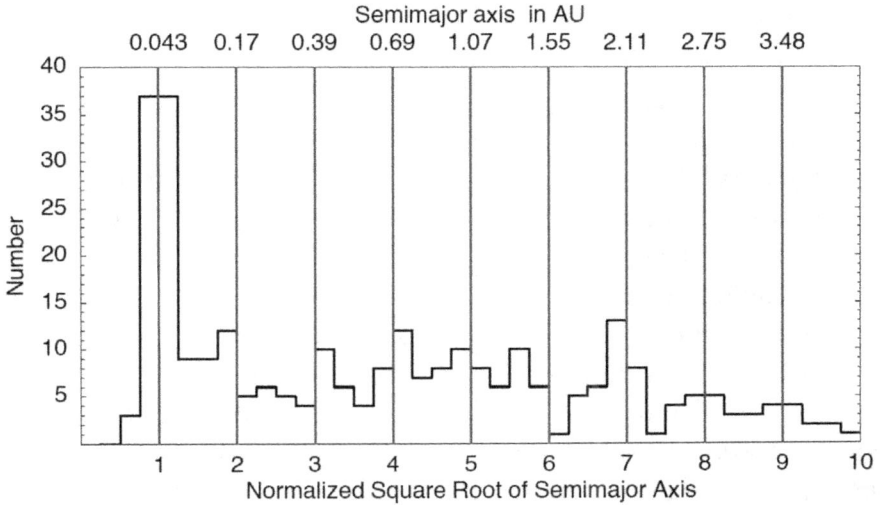

Figure 33 *Observed distribution of exoplanets (2008).* the fundamental orbital peak is dominant, and secondary probability peaks are observed at the expected values, around the positions of the solar system planets.

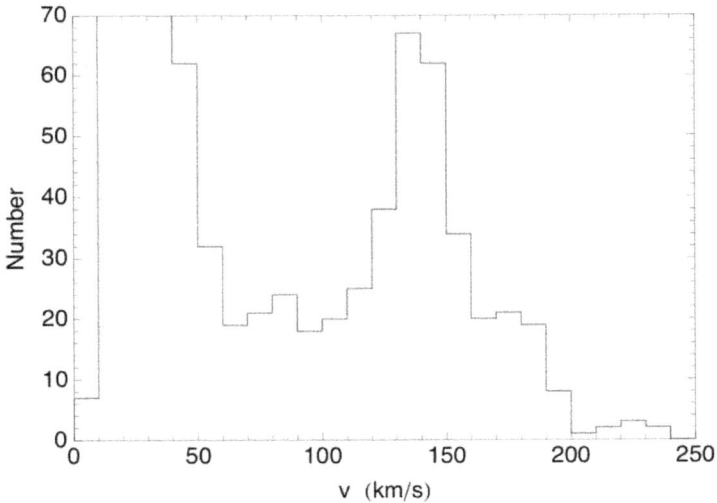

Figure 34 *Observed distribution of velocities for 900 exoplanets with masses larger than one twentieth of the mass of Jupiter (2018).* The expected probability peak around 150 km/s is clearly apparent.

Planets Around a Pulsar

More extraordinary yet is the case of the three planets discovered around the pulsar PSR B1257 + 12 by A. Wolszczan and his collaborators. The agreement with theory is so precise that the second-order terms can be tested.[195] In fact, the conservation of energy and of the center of gravity of the sub-disks finally giving rise to the planets after accretion implies that, in the absence of perturbation, the planets should be found at the average values of the probability distributions (given by n^2 + $n/2$), instead of the peaks of these probability distributions (in n^2). The agreement was (in 1998) at the level of around 1/10,000 (see figure 35), ten times better than with the formula in n^2 (which would already have been remarkable).

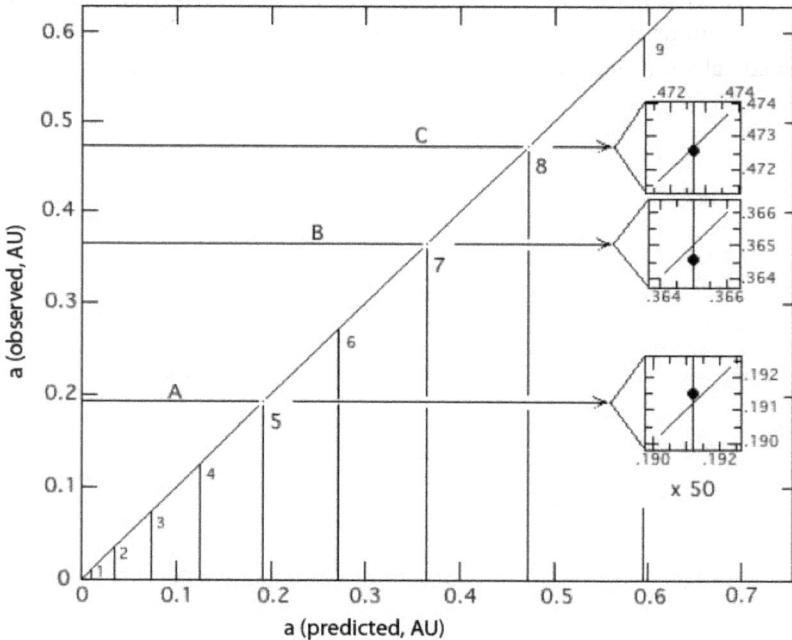

Figure 35 *Fit between theory and observations for the system of three planets around pulsar PSR B1257 + 12.* The diagram compares the observed values of the semimajor axes of three planets orbiting around the pulsar PSR B1257 + 12 to the possible values theoretically predicted. The agreement between prediction and observation is much more precise than the resolution of the

diagram, so three insets enlarged by a factor of ≈ 50 have been inserted to show the slight residual differences. We have supposed here that the mass of the pulsar was 1.48 times the solar mass, which corresponds to a velocity in the fundamental state for this system of 3 w_0 (which is just the Keplerian velocity at the Sun radius). The quality of the alignment does not depend on this choice, only the value of the slope. The probability of obtaining such a configuration by chance would be on the order of 1/100,000.

This agreement has since still been improved. By analytically integrating the motion of these planets to about 1 billion years (since their formation), we have been able to show that the conditions which held in their formative epoch were still achieved today, without major perturbation. This is due in particular to the small masses of the three planets (on the order of 3 times the Earth mass and the Moon mass). Moreover, the continuing observation of the system over more than 13 years by Wolszczan, and the account of mutual gravitational effects between the planets has allowed for a great improvement in the determination of their orbital elements. These new values could have degraded the agreement with the scale relativity theory predictions. On the contrary, this agreement has been improved by a factor 10, reaching a precision on the order of some 10^{-5}. Such a precision is usually reserved for relative celestial mechanics measurements, while it is applied here to the position of the planets themselves (their semimajor axes, and also the eccentricities).[196] To give a comparison, this means that if one makes a model of this system at a scale of 100 m (this has been in process at Paris-Meudon Observatory), the positions of the planets would fall on the expected law with a precision of some millimeters.

Structure Formation

In the standard theory of gravitational structure formation, whatever the scale, structures cannot form from an initial constant density. One needs an initial fluctuation, which is difficult to reconcile with the conditions of the primeval universe, which were remarkably smooth. In the scale relativity/macroscopic Schrödinger approach, this problem is solved: indeed, a constant density is expressed as a harmonic oscillator potential, and the Schrödinger equation has well-known stationary solutions in that case (see Fig. 36).[197]

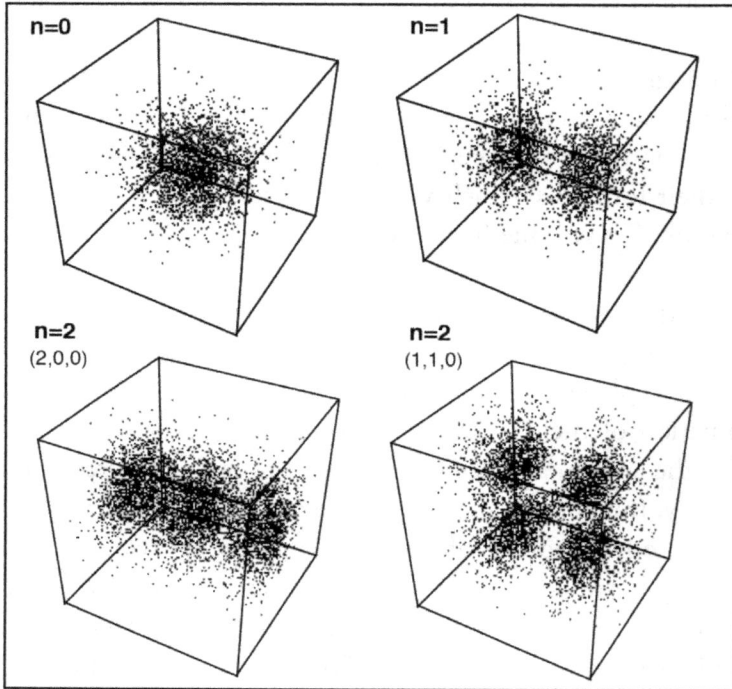

Figure 36 *The first three modes of the solution of a Schrödinger equation for a particle in a 3-dimensional harmonic oscillator potential correspond to the gravitational potential of a background of constant density* (the mode n = 2 decays into two submodes). In the scale relativity approach, the geodesic equation can be integrated in terms of a Schrödinger equation, so that structures are formed even in a medium of strictly constant density. Depending on the value of the energy, discretized stationary solutions are found that describe the formation of one object (n = 0), two objects (n = 1), etc. I have simulated these solutions by distributing points according to the probability density. The mode n = 1 corresponds to the formation of binary objects (binary stars, double galaxies, binary clusters of galaxies, etc.).

The remarkable result is that these solutions are spontaneously self-organized in terms of one object, a pair of objects, an alignment of objects (three or more), or a trapezoidal shape of four objects, etc. This explains a well-known fact of astronomy: stars, clusters of stars, galaxies, and clusters of galaxies appear as single objects, but also very frequently as pairs, of stars—more than 50% of stars in the galaxy are double—but also pairs of star clusters, of galaxies, and of clusters of galaxies. Concerning the alignment and quadrilateral shapes, it is clear that they cannot be stable because of the mutual effect between the bodies. Such shapes can therefore appear only at the epoch of their formation, than are expected to change. Here again, this is confirmed by observations: the zones of star formation show in a systematic way star alignment and trapezia (such as the Orion trapezium cluster, well-known to amateur astronomers). The same is true for galaxies: compact groups of galaxies showing similar shapes do exist and are considered as young systems seen just after their formation. The reason for these shapes is not known in the standard framework, while it is clearly predicted and understood in the scale relativity theory.

Planetary Nebulae

Planetary Nebulae are, despite their misleading name, stars which have ejected their outer envelope. Besides the naturally expected spherical shape, they have shown a wide diversity of different shapes which are not really explained by the standard theory. The scale relativity approach has been able to account for them in detail, see Fig. 37 (predicting in particular the numerical values of the ejection angles). Moreover, some new structures which had not yet been observed at the date of this prediction (for example, the bottom-left shape of Fig. 37) have subsequently been discovered.[198]

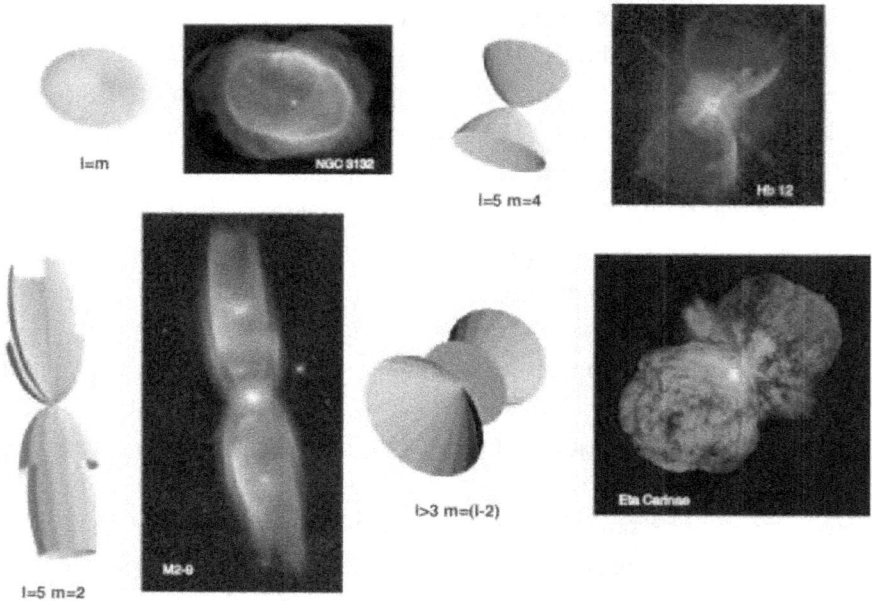

Figure 37 *Comparison between quantized shapes predicted from the scale relativity equation* (taking the form of a macroscopic Schrödinger equation) for the ejection of gas from a center, and typical observed Planetary Nebulae (outer shells ejected by some stars).

Galaxies

Several structures are also observed at the extragalactic scales, which demonstrate the expected universality of structuring in velocity space (in agreement with, and as a consequence of, Einstein's equivalence principle). For example, the rotation velocity of spiral galaxies is known to increase from their center and to show a flat maximum in their outer regions: this is one of the arguments for a missing component in the dynamics of extragalactic objects, sometimes attributed to "dark matter," since the rotation velocity should decrease outside the galaxy according to Kepler's law. As can be seen in Fig. 38, the probability distribution of these maximum velocities shows a well-defined peak at the same value (around 150 km/s) as the fundamental orbital of our inner solar system, which corresponds also to the main exoplanet probability peak.[199]

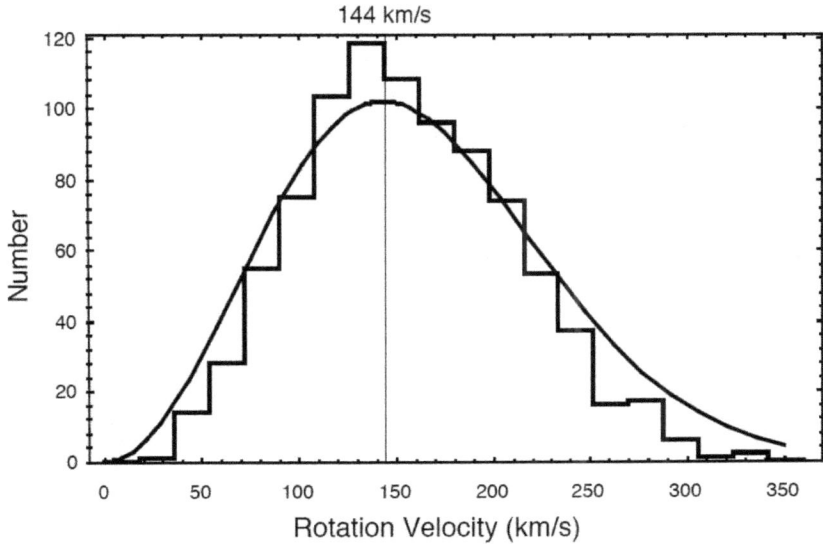

Figure 38 *Distribution of the outermost observed velocities in spiral galaxies* (flat rotation curves) from the catalog of rotation curves for 967 spiral galaxies by Persic and Salucci.

A similar result is obtained for galaxy pairs. In 1977, a statistical effect of quantization of redshift between the two members of pairs of galaxies was suggested by William Tifft of the University of Arizona.[200] Such an effect seemed very improbable, since the redshift is expected to come from radial velocity, i.e. from the velocity projected along the line of sight, while the pairs can have any orientation. Indeed, it has not been confirmed with modern data. The catalogs of galaxy pairs were very scarce at that time, and the numbers therefore showed strong relative fluctuations. Today, new catalogs have been built of more than 13,000 galaxy pairs, and the distribution of their observed radial (i.e., projected) velocity differences is quite smooth and monotonic as expected, in agreement with the standard view of the redshift as radial velocity. However, if one performs a statistical deprojection of these velocity differences, one finds that some values of the three dimensional velocity are more probable than others (see Fig. 39).[201] Moreover, one obtains a remarkable new result. Pairs of galaxies, from the viewpoint of dynamics, correspond (in what one calls reduced coordinates and masses) to exactly the same "Kepler" gravitational problem as a star-planet couple, but at a mass scale and distance scale larger by a factor 10^{12}. The universality of the scale relativity approach, despite this huge scale ratio, leads one to expect to recover the same fundamental level solution as for our solar system and exoplanets as the main probability peak at around 150 km/s: this is exactly what is observed, in a systematic way, on all the pair catalogs studied.

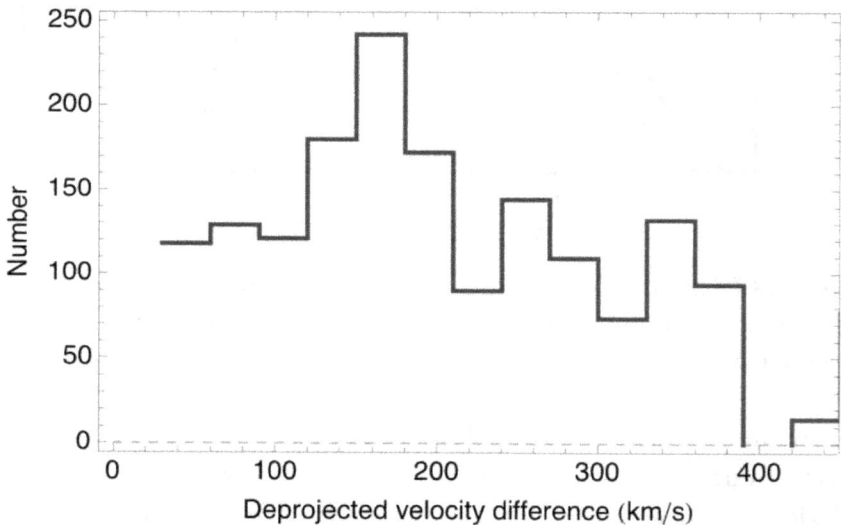

Figure 39 *Histogram of the distribution of the velocity differences between the members of galaxy pairs* (13,114 pairs in the Nottale-Chamaraux catalog), obtained from a statistical deprojection of the radial velocities. The main probability peak lies around 150 km/s, the same as for exoplanets (see Fig. 34), as expected from the universality of the scale relativity approach to formation and evolution of gravitational structures.

Let us note, finally, that these results are only part of a much larger set of structures predicted by the scale relativity theory. Toward extremely large and small scales (temporally and spatially), other more definite consequences of scale relativity are expected.

In the case of gravitational systems, only the Kepler problem has been discussed here, but other types of quantization are predicted and have been able to be confirmed for other potentials, the extragalactic case (the universe at large scale) being particularly interesting. Verifications of the theory in multiple systems have been obtained, over a vast range of scales: the innermost solar system, satellites of giant planets,[202] asteroids and comets, binary stars,[203] zones of formation of stars, planetary nebulae as we have seen, galactic dynamics, local group of galaxies,[204] dynamics of groups and clusters of galaxies, etc.[205]

Other Sciences

The theory of scale relativity/fractal spacetime has also been successfully applied to sciences other than physics. This is not unexpected, since many realms show an explicit dependence on phenomena over scales, as in geography (which is even the origin of the concept through the scale of a map), or living systems (biology). There are essentially two kinds of possible applications.

One of them is the use of new scale laws to account for misunderstood scaling phenomena. Recall that the scale relativity method amounts to describing the scale dependence in terms of differential equations acting in scale space. These differential equations, as in the case of motion laws, are obtained from an optimization principle that ultimately comes back to the principle of relativity itself. This allowed us to generalize the simple self-similar fractal laws (of constant fractal dimensions) to more complex laws: it is similar to the jump from inertial laws to Newton's dynamics in the case of motion laws.

The other application consists of applying the macroquantum Schrödinger-type approach to systems for which the three underlying conditions (infinity of virtual trajectories, fractality, and two-valuedness of derivatives due to local irreversibility) are fulfilled, at least as an approximation and on a large enough range of scales, which corresponds to many turbulent or chaotic systems. We have already seen many applications of this kind to gravitational structuring in astronomy.

Let us briefly review (in a non-exhaustive way) some of the new applications of scale relativity to other sciences and domains.[206] Note that scale relativity has led to important mathematical developments, in particular by J. Cresson and F. Ben Adda;[207] to deep studies in philosophy of science, especially led by Charles Alunni,[208] and in sociology by V. Bontems and Y. Gingras;[209] to applications to psychoanalysis and neurosciences in collaboration with Pierre Timar;[210] and to proposals, in collaboration with Thierry Lehner, of new technological devices using macroscopic quantum potentials.[211]

One of the most promising applications of the new scale laws has been in geography. With Philippe Martin and Maxime Forriez, we showed that the simple self-similar fractal laws with constant fractal dimension were insufficient for describing drainage basins, while an approach in terms of varying fractal dimension (corresponding to a kind of "scale dynamics") allowed a very precise and predictive description (see Fig. 40).[212]

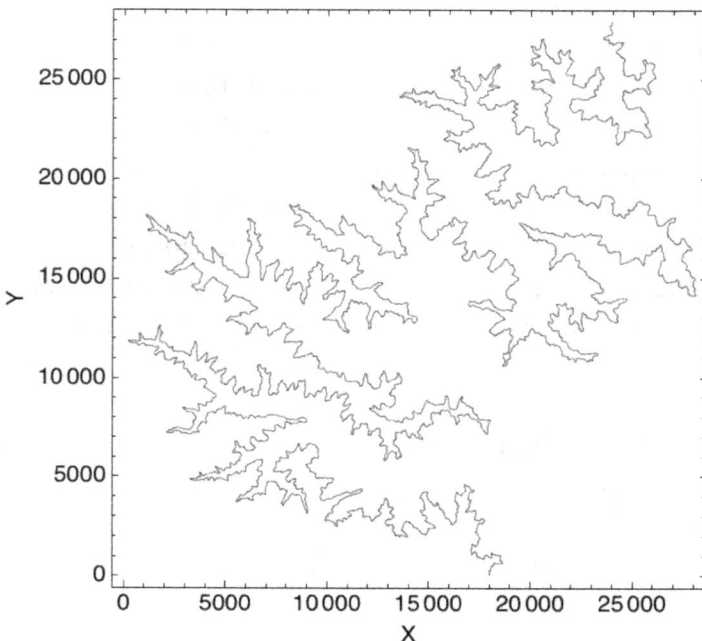

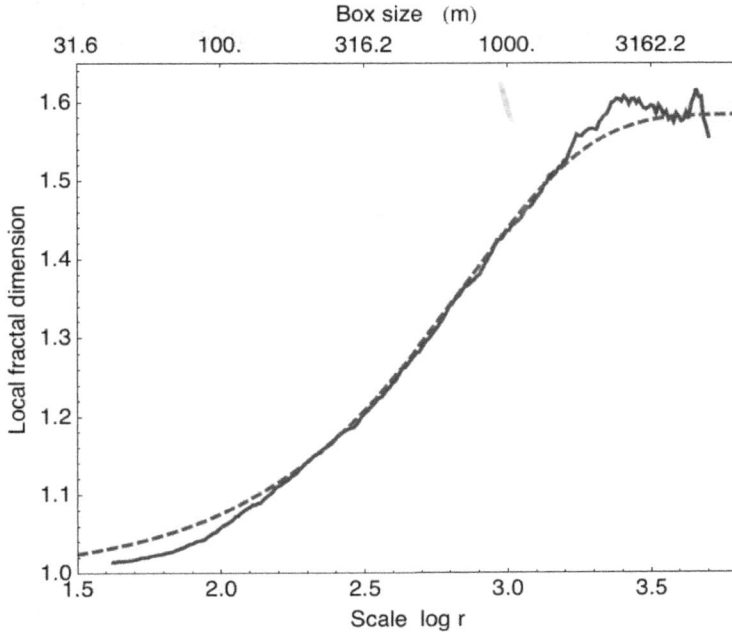

Figure 40 *Variable fractal dimension in geography.* The first figure is a contour line in the drainage basin of the Gardons (France). The second figure shows the observed variation with scale of its fractal dimension (blue curve). It is compared with its variation (red dashed curve) theoretically expected from an analysis of the fractal organization of the rivers and of their corresponding drainage basin. The maximal fractal dimension is also theoretically predicted by the same analysis to be log3/log2 ≈ 1.585, in fair agreement with its observed value.

In geosciences, we applied, as early as 2006, the scale acceleration laws obtained as solutions of scale differential equations to the question of the arctic sea ice melting.[213] This acceleration of the melting was understood from the fractal structure of the ice fracturation. At that time, the current view was that the sea ice would not melt fully (after summer) before about the year 2100. We were the first to use a critical scale law (that is, a law including in its expression an explicit limitation in time) to show, before the observation of the surprisingly low 2007 value, that the melting was accelerating and that the arctic could be

257

void of ice in the September months as early as 2030 or before. This is now a well-known fact described by many different models (see Fig. 41).

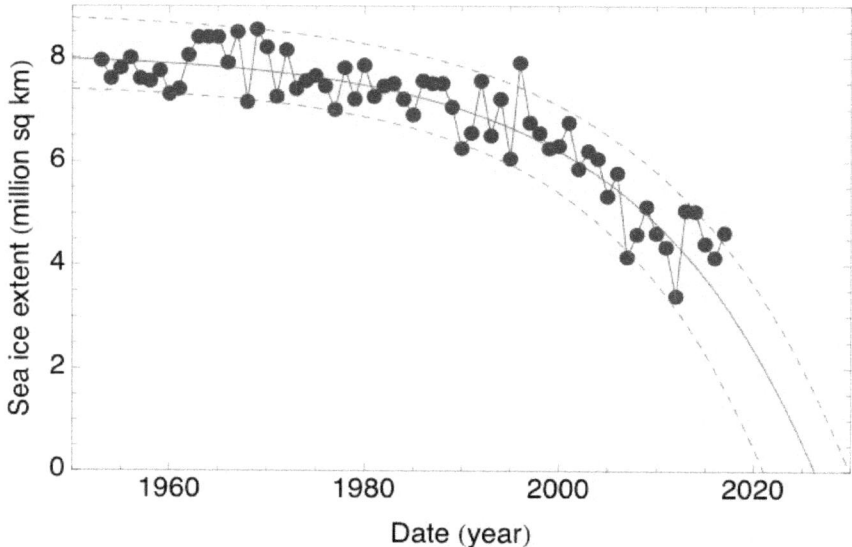

Figure 41 *Comparison of the observed sea ice extent in the arctic with an exponential model of melting acceleration (2017 data).* The full melting (just after summer) is expected between 2020 and 2030. Accounting for the ice thickness, one finds a date even earlier than 2022 according to this model.

In biology, we developed models of morphogenesis and self-organization in collaboration with Marc Pocard and Etienne Rouleau, then with Charles Auffray and Eric Eveno in systems biology. More recently, this approach has been applied in the different context of plant growth in collaboration with Philip Turner. I shall just briefly illustrate this approach by two examples.

The first example concerns morphogenesis of plant-like structures. Based on the underlying fractal substratum of living systems (and on their multi-scale organization), one can suggest that their growth equation, written as a geodesics equation, takes a macroscopic Schrödinger form on account of the fractal geometry at a large range of scales. The most simple case to look at is the free growth of a system from a center. One obtains probability density distributions depending on the ejection angles, similar to the planetary nebulae case, but here along only one direction instead of two. We have represented in Fig. 42 one of the results obtained by "injecting" matter along the peaks of probability. We clearly get flower-like structures. This is just the most probable shape, which is accompanied by many slightly different manifestations resembling it, but with different angles. In the figure, we have changed a tension term, which allows the opening of the "flower," and we have added the effect of gravity. The number of petals and the number of shells are controlled by integer quantum numbers. This is not contradictory with the genetic determination of biological shapes: it supports it and simplifies it, since the genetic code would only need to store these integer numbers to determine the shape.

Figure 42 *This is not a flower.* Opening of a flower-like structure, solution of a macroscopic Schrödinger equation describing growth from a center. The various elements of the shape (petals, sepals, stem, pistil) are all obtained once from the same solution (it is not a flower model).

Another example of application concerns cell organization and division. One of the most crucial questions in life science is the origin of cells. The first living organism was a prokaryotic cell, and the most complicated present organisms are still made of cells. They look like a kind of "life atom" or "life quantum." In the scale relativity framework, this is easily understood through the natural quantization induced by the Schrödinger form taken by the equation of dynamics. Moreover, among the new scale laws obtained from scale differential equations, some of them (linked to contraction/dilation) yield natural membrane or cell wall structures. Another basic element of life is the process of cell division. In a classical system, the energy can change in a continuous way, and if one increases the energy, one generally increases the size of the system. But in a Schrödinger regime, there are only some quantized values of the energy for which structures can persist, and the effect of jumping from the fundamental state to the first excited state does not yield an increase of size, but a spontaneous division into two structures of about the same size as the original one.

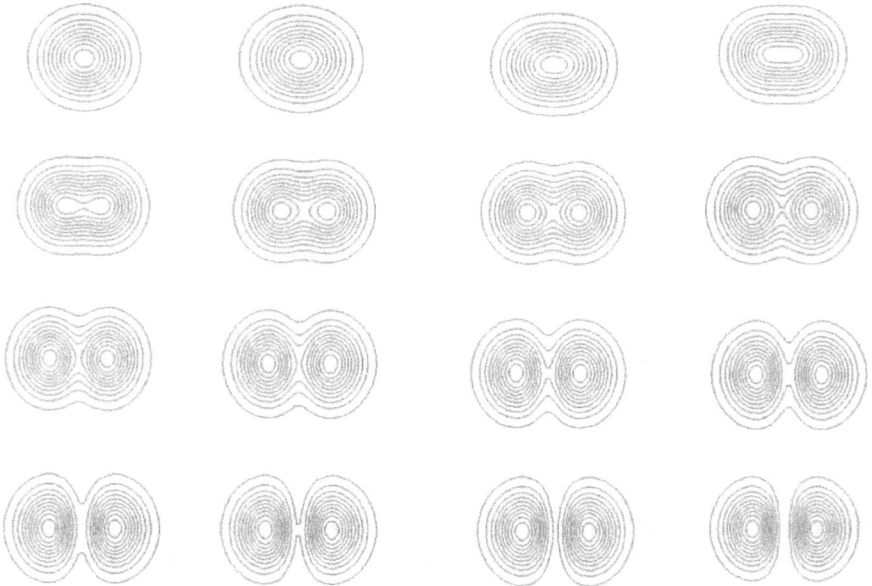

Figure 43 *Successive steps of a spontaneous "cell-like" division.* The first (top left) and last (bottom right) structures are the fundamental level and first excited state of the stationary Schrödinger equation. They correspond respectively to one body and two separated bodies: one jumps from one state

to the other by simply changing the (quantized) energy. The intermediary steps are also solutions, but of the time-dependent Schrödinger equation, and can therefore only be transitory.

It is clear that a real cell and real cell division are far more complicated than this simple description. The idea is simply that the division process may correspond to a very fundamental and basic law which has been implemented by biological evolution (in combination with many other elements, some of which are also accounted for by the scale relativity approach).[214]

Let us end this brief review of applications of scale relativity by recalling various occurrences of log-periodic laws, in particular in life sciences (species evolution and embryogenesis), history (evolution of civilizations), and geosciences (earthquakes). These scale laws are characterized by fluctuations around self-similar laws accelerating toward a critical date or decelerating from this date. They can be obtained as a natural consequence of scale covariance.

In collaboration with Pierre Grou and Jean Chaline, the major evolutionary leaps of several lineages of species evolution have been shown to be accelerating toward a critical time, specific to the lineage, while some of them, like echinoderms, decelerate from their date of advent (see fig. 44).[215] The critical date of accelerating lineages can be interpreted as the end of their capacity of evolution, not their extinction *a priori*. For example, for some dinosaurs the critical date is about 140 million years in the past, while they disappeared 65 million years ago. It is just that the morphological criterion used (shape of the legs) ceased to evolve after the critical time and remained adapted to the environment. In the case of rodents, which show the largest variability among mammals, the critical date can be as large as 60 million years in the future. For primates (including hominidae), T_c is two millions years in the future, but all this research (which is concerned with lineages, not species and even less individual beings) is at the scale of resolution of million years, so that this may mean that the critical date has been reached for our lineage. Note that in this case, a kind of retrodiction has been gained, since the fit of the jump dates to a log-periodic law has revealed a missing link (from the viewpoint of fossil records) about 10 million years in the past. This is exactly the estimate from genetic distances (which uses the "distances," i.e. the amount of difference, between two DNAs) of the date of the pangorilla-homo common ancestor, whose fossil has not yet been found.

In collaboration with Roland Cash, the same kind of law has been used to analyze the chronology of the main dates of embryogenesis and human development. We have found a clear decelerating log-periodicity beginning right at the conception date and including events like the birth and the passage to adulthood.[216]

Under the impetus of the economist Pierre Grou, the same kind of analysis has been also applied to the evolution of societies/civilizations, to economic history, and by Ivan Brissaud to many examples of the chronology of technological innovation.[217] One can show that the alternations of crises and economic upturns that most societies experience can be described by log-periodic scale laws accelerating toward a critical date (dependent on the civilization studied). In particular, we have found (in 1996) that the whole western civilization shows an acceleration toward a critical date around 2050-2080, since the neolithic and even before (see Fig. 46). This result has subsequently been confirmed and reinforced by Johansen and Sornette, using different economic indices. This "super-crisis" should not be considered as catastrophic in itself. It is just the manifestation of a natural change, of an inevitable transition, like the teenage crisis which is the passage to adulthood.[218] Then the critical date does not mean that something specific will happen at this precise date: it is more like the summit of a mountain. Actually, it has now become clear to everyone that humanity is entering a crisis (or natural change) period of a completely different level from previous crises. The question is: shall we accept it and participate in this inevitable change, allowing a soft transition, or deny it and fight against it, which can only lead to suffering.

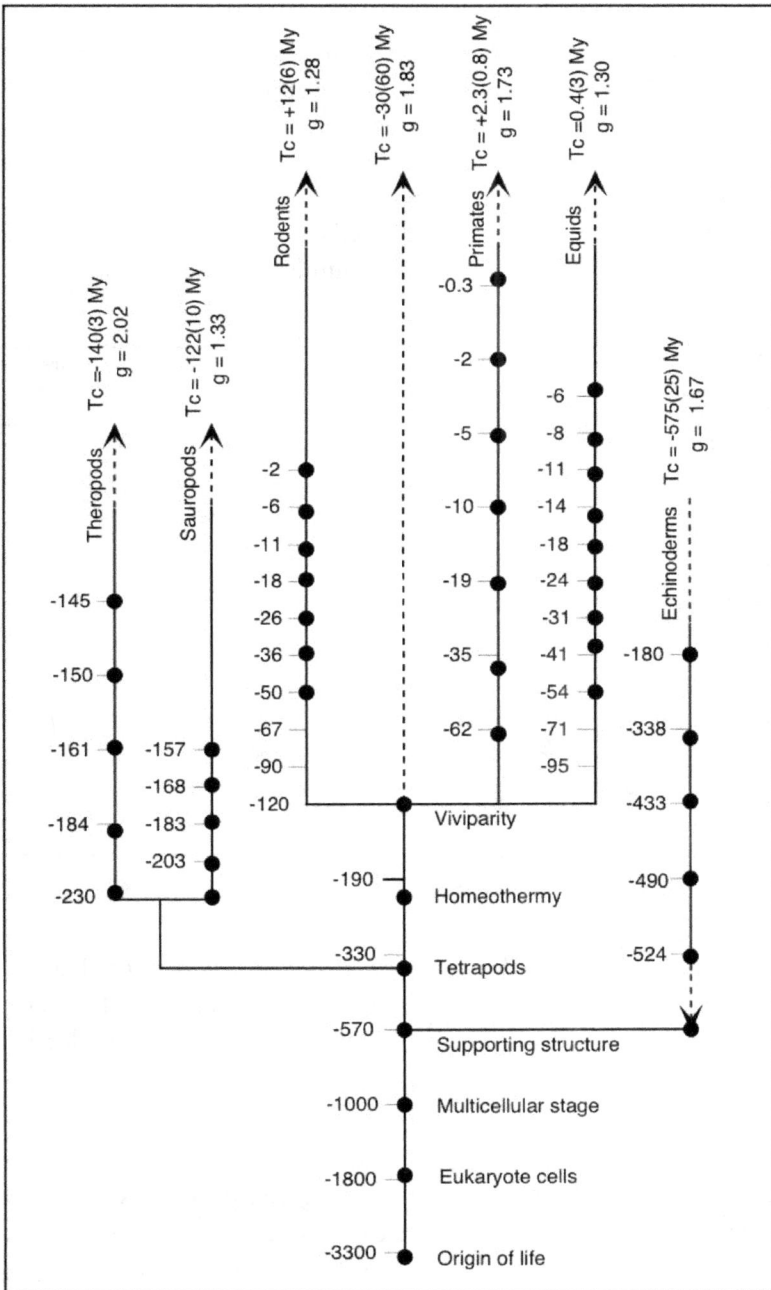

Figure 44 *Log-periodic evolution on the "tree of life."* The dates of major evolutionary events of seven lineages (common evolution from life origin to

viviparity, Theropod and Sauropod dinosaurs, Rodents, Equidae, Primates including Hominidae, and Echinoderms) are plotted as black points in terms of log(T_c - T). Their log-periodicity therefore appears as a quasi-periodicity in this diagram, although they strongly accelerate or decelerate in real time (see the numbers, given in million years in the past). They are compared with the numerical values from their corresponding log-periodic models (computed with their best-fit parameters). The adjusted critical times T_c and scale ratios g are indicated for each lineage. The number between brackets is the uncertainty on T_c in million years.

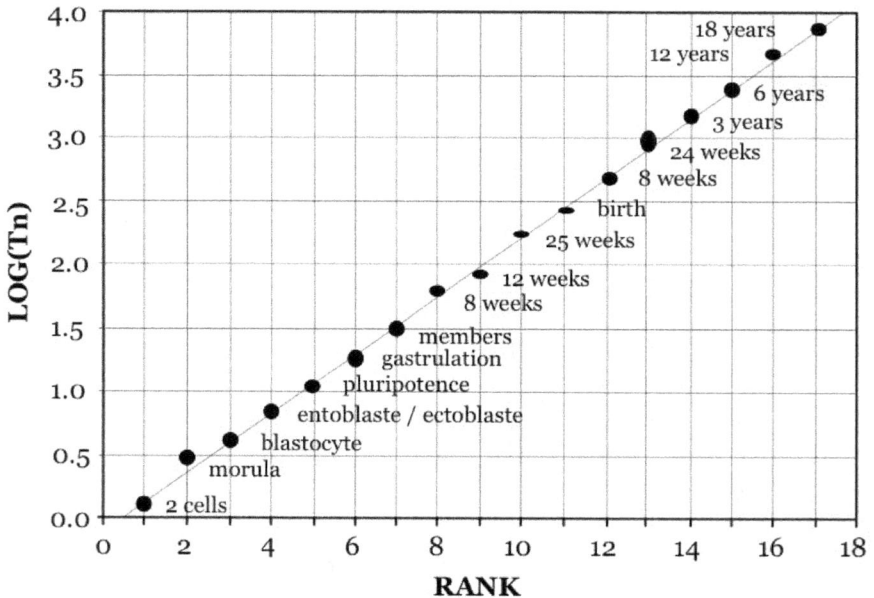

Figure 45 *Main steps of embryogenesis and human development.* The decimal logarithm log(T_n) of the dates (in days), counted starting from the conception date, is plotted in terms of their rank, showing a log-periodic deceleration with a scale ratio g = 1.71±0.01.

Let us end with an application in geosciences, performed in collaboration with Fred Heliodore, using log-periodic laws to study earthquakes. This has been one of the first domains of application of log-periodicity, by Didier Sornette and his collaborators.[219] However, here the idea is different, since, instead of searching for log-periodic variations of external measured quantities (which are not always available), we directly use the observed magnitudes and rates of earthquakes. The large seisms are systematically followed by aftershocks: the rate of these aftershocks shows peaks at some particular dates which decelerate in a very clear log-periodic way from the main earthquake date (see Fig. 47). The aftershocks also show bursts of their magnitudes, the burst dates being also governed by the same log-periodic law.[220] There is therefore some predictability of the aftershocks, but with uncertainties which increase with time, due to the logarithmic nature of the law.

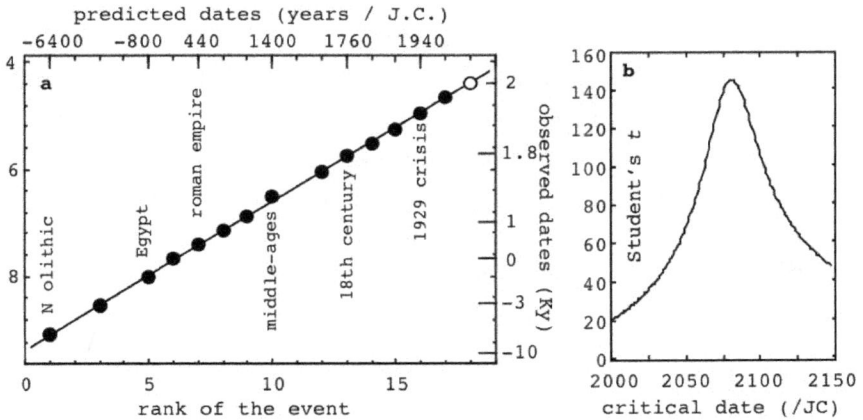

Figure 46 *Comparison of the median dates of the main economic "crises" of Western civilization with a log-periodic accelerating law of critical date T_c around 2075 and scale ratio $g = 1.32$ (figure a).* The last white point corresponds to the predicted next crisis predicted around 2000 at the date of the study (1996), as has been later supported in particular by the 1998 and 2000 market crashes. Figure b shows the estimation of the critical date. This result is statistically highly significant (the probability to obtain such a high peak by chance is smaller than 1/10,000).

Finally, one of the most extraordinary applications of the theory of scale relativity will perhaps be to understanding turbulence. Although it was described by Leonardo Da Vinci several centuries ago, and though the equations of fluid mechanics date back to the beginning of the 19th century with Euler, Navier, and Stokes, turbulence remains essentially misunderstood. The breakthrough came from Louis de Montera's insight, according to which the scale relativity approach and methods should be applied, in the description of a turbulent fluid, in *velocity space* instead of position space.[221] As a consequence, the equations of dynamics (or more precisely, their derivative) can be given the form of a macroscopic Schrödinger equation. In collaboration with Thierry Lehner, definite proofs of the validity of this proposal have been obtained from an analysis of experimental data, leading in particular to an understanding of the intermittent character of turbulence.[222] Therefore, fluid turbulence implements my now 25-year-old insight[223] according to which a fractal medium (the turbulent fluid in velocity space) may play the role of a fractal and nondifferentiable space, and then confer to the particles which move in it (the test particles and/or the fluid particles themselves) macroscopic quantum-type properties. This means that, at last, not only observational, but now experimental evidence has been obtained in laboratory of the existence of a macroquantum regime governed by a Schrödinger equation in a classical system, which was a theoretical prediction of the scale relativity theory.

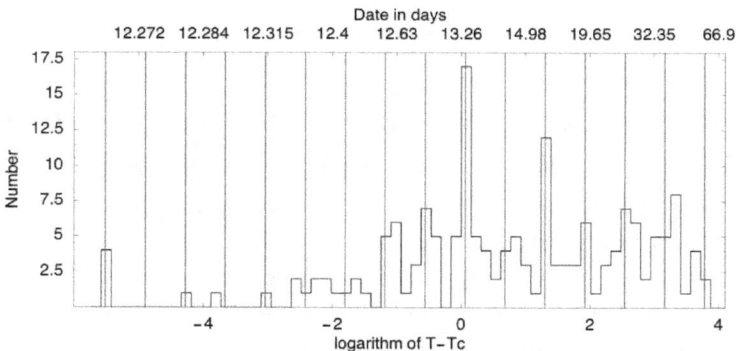

Figure 47 *The rate of aftershocks which followed the main Sichuan earthquake of May 12.27, 2008 (magnitude 7.9) is shown in function of the logarithm of the difference between the aftershock date and the main earthquake date, ln(T − T$_c$), for Tc = 12.2665 May 2008.* The vertical lines indicate the peaks expected according to the best fit period, 0.619 (corresponding to a scale ratio g = 1.86). It is clearly seen that the observed peaks of the aftershock rate agree very closely with a log-periodic law decelerating from the main seism date, from time-scales of hours to several months.

CONCLUSION

In this work, I have endeavored to show how the principle of relativity is not a definitively fixed principle, but a developing one, and to present the outline of an attempt to generalize it. I have tried to describe here the underlying physical principles of the new theory, and have also tried to present (in a non-exhaustive manner) some possible consequences of this new approach. As we have seen, the theory of scale relativity enables us to propose concrete solutions to a certain number of problems which have remained unresolved in the framework of current physics.

However, scale relativity, being in addition an extension of the relativist way of thinking, allows us to shed a new light on questions that go much further than physics alone. Many apparently unsolvable problems find a solution by the mere extension of our mental structures. What seemed impossible at one point is no longer so one day, not because the outside world has changed, but because our mind has opened to new possibilities, once hidden. As a way of concluding (and of beginning, since we are only at the start of such an enterprise: it is not impossible that it will take several centuries to decipher the abundance of structures entailed by the abandoning of the hypothesis of differentiability), we will discuss some of these questions.

By explicitly introducing scales into our way of thinking, we can better understand certain problems coming from scientific fields other than physics. Thus, in the life sciences, there is a tendency to reduce the description of living systems to their genetic code or their genes. Yet these systems, especially the most highly evolved, are characterized by the nesting of different levels of organization: atoms, genetic code bases, DNA, chromosomes, cell nuclei, cells, tissues, organs, organisms. It is possible to envision that all these levels coexist, that each one has its own way of functioning, that none are reducible to the others, but that they are all connected to one another. Each level is essential, and contains new information that cannot be reduced to that of the preceding levels. In addition, this information circulates between scales, from the small to the large, but also in the other direction, assuring the cohesiveness of all the levels.

Over the passing of billions of years, from unicellular to multicellular life, new complex structures do not replace older ones: the multicellular being that we are is a set of cells, not a single enormous cell. It is a new level of organization which has appeared. Not only does it not suppress the cellular level, but on the contrary it protects and expands its possibilities of surviving.

This logic can be extended to social organization. We can live at the level of local associations, of the city, of the region, of the nation, of the continent, and of the planet: none of these levels should substitute another one; all are necessary, being able to have their own mode of democratic organization, and assuring the cohesion of everything by a double communication, between elements of the same level (for example the European countries among themselves), but also between different scales (with Europe as a whole and with their separate states).[224]

One question which has fascinated physicists and philosophers is that of the nature of time and its passing. Relativity enables us to elucidate several aspects.

First of all, the special relativity of Einstein and Poincaré considers time as the fourth dimension of spacetime. While we are conscious of the three dimensions of space, our consciousness of time escapes us—more specifically, we are not conscious of time directly, but only through a change in space—, but this is purely a consequence of our situations. As we are able, for the moment, to travel only at slow velocities relative to the speed of light, the rotations which are accessible to us in spacetime are minuscule, while we can turn as we like in space (recall that a motion in space is but a rotation in spacetime). If one day humanity is able to access velocities close to that of light, we would "see" time by experiencing four-dimensional rotations.[225]

Special relativity also implies, by the existence of light cones, the separation between past and future, and thus allows causality. Had spacetime been Euclidean, it would have been possible to make a complete turn in spacetime and to return to the past simply by accelerating, with the problems of causality that this would have provoked. The existence of a maximal speed that cannot be exceeded (the speed of light) limits the possible rotations and separates the past and the future.

Let us now consider the question of the passage of time. Why does time pass? We have learned that time is one of the coordinates of spacetime and that what is time in one reference system can become space in another. In addition, this question has no meaning in itself. One should ask whether space passes as well! But the "passage" of space is motion. It is in the nature of things that, with respect to whatever reference system, the three spatial coordinates change. It is immobility which poses a question, not change. Even if we believe we can speak of rest for a macroscopic body, it is never anything but an approximation, and quantum mechanics teaches us that the particles which it is made of are never at rest. Rest, ultimately, does not exist between two objects: even for a temperature of absolute zero, which is an unattainable horizon, one would have to take into consideration the energy of a vacuum. There is only rest in an extremely particular reference system, which is the proper reference system specific to each object, moving as the object moves.

We can now think, no longer in terms of space over the course of time, but in terms of spacetime. The passing of the temporal variable then poses no problem, quite the contrary. The other three (spatial) coordinates always being subject to change, it is the same way with the fourth (temporal). "Change is like nothing," in space as well as time. Nevertheless, rest in space does indeed exist, in the reference system proper to each object. What is it for time? When we speak of a proper reference system, we mean a reference that moves with the object, but in which time passes (this is proper time, which defines the invariant of Einstein's theories). But we can go further, and place oneself in *a reference system completely proper in spacetime*, which is quite simply the reference of the *present*. In this reference system of the *here* and *now*, which is finally that of each individual object, time no longer passes, and the relativistic invariant takes the form that it has for light, $ds = 0$! It is a completely free and empty reference. But certainly the extension and the multiplicity of objects implies the coexistence of multiple different reference systems so well that, in the changes of reference, the passage of time and different phenomena reappear.

If the passage of time can thus be understood, what about its arrow? Why does it always seem to flow in the same direction for all objects? This manner of asking the question corresponds, once again, to the limited vision of our current experience of reality. It is true for us, macroscopic bodies conscious of the passage of time at the resolution of a hundredth of a second at best, and moving at low speeds. However, we can change the scale and the rate of movement, and imagine ourselves to be elementary particles, traveling at speeds close to that of light, and being able to discern temporal resolutions of 10^{-25} seconds. Then what would we see around us? A flood of particles, some of which would advance in the direction of time, but others which would *move backward in time*. Feynman has explained antiparticles to be particles which travel backward in time. The more we progress toward smaller scales, the more the number of particles and antiparticles are balanced due to the creation and annihilation of pairs. Toward very small temporal resolutions, meaning at very high energies, there are as many objects that travel backward through time as objects which move forward. Furthermore, according to the description in terms of fractal trajectories within spacetime, each individual particle oscillates between past and future at very small scales, as Garnet Ord showed at the beginning of the 1980s. Time's arrow is also dependent on scale.

One final question concerning time is that of irreversibility. It is not the irreversibility or the reversibility of the laws in themselves which pose a problem, but the coexistence of the two: one of the biggest problems in physics is precisely the apparent contradiction between the reversibility of fundamental laws and the irreversibility of real macroscopic systems.[226]

Scale relativity will not claim that the "true" laws are reversible— or irreversible, but states the question otherwise: reversibility and irreversibility are no longer absolute terms, but become relative on the scale under consideration.

First of all, it introduces a local level of description, more profound than that of current quantum mechanics. At this level, the fundamental laws are irreversible: there is no invariance by reversal of direction of the infinitesimal temporal element.

The methods of quantum mechanics are then reconstructed by combining the two directions of the passage of time: the laws obtained, at a more global level, become reversible again, in terms of a new descriptive tool (the wave function), which accounts for both time directions together using complex numbers.

Next, the reduction (also called "collapse") of the wave function reintroduces a fundamental irreversibility tied to measurement in the passage from quantum to classical. Still, the laws of classical mechanics for point particles are reversible (this last transition remains problematic).

The passage to statistical mechanics for a very large number of particles reintroduces, with thermodynamics, a fundamental irreversibility described by the law of increasing entropy. With dynamical chaos, this becomes true even for a small number of bodies: the laws of celestial mechanics become chaotic starting from the three-body problem. One then finds irreversibility again over very large scales of time, beyond the horizon of predictability implied by chaos.

Finally, we have seen in the fifth part that one can reconstruct on this basis a quasi-quantum macroscopic theory, in which reversibility again occurs, no longer at the level of individual trajectories, but at the level of the structures considered in their globality.

Thus the laws of physics can go through different "phase transitions" in the changes of scale, and move from reversibility to irreversibility and vice versa.

Before closing, let us add a few words on the problem of zero and infinity. Kant, wishing to study the problem of the limits of the world in time and space, was led to one of his antinomies of pure reason. Based on the impossibility of an infinite series of phenomena and the necessity of stopping somewhere, the thesis states that "the world has a beginning in time, and is also limited in regard to space." But the antithesis does not see why one should stop here rather than there, and states that "the world has no beginning, and no limits in space, but is, in relation to both time and space, infinite." Kant's goal was not that "of finally deciding in favour of either side, but to discover whether the object of the struggle is not a mere illusion."[227] Scale relativity's introduction of finite and unsurpassable scales of resolutions, which nevertheless possess the physical properties of zero and infinity, finally resolves, it seems to me, this contradiction.

But one must not forget that this solution is obtained at the cost of a radical change of the form of the laws of dilation toward the infinitely small and toward the infinitely large. The nature of the Planck spacetime scale and the cosmic scale become almost unthinkable to us. There are no longer any laws at these two extremities of the world of scales, where physics becomes totally degenerate. It would take a "microscope" capable of infinite enlargement to "see" the Planck length. Inversely, the measure at the Planck resolution of any interval of length, even that of the whole universe, would be—the Planck length itself! Absurd statement? No, since this measurement would necessitate an infinite energy, which simply cannot be found except by taking the universe in its entirety. The Planck scale contains the whole universe! At the other extremity, if one considers the world at the maximal resolution, assumed to be given by the cosmological constant, and one wishes to double this resolution, nothing happens: the resolution has not doubled and the measured contents have not changed. The existence of an invariant cosmic scale implies the statement, finally coherent, that the universe is "one," seen at its own resolution. Zero and infinity meet. The physics of fractal spacetime is the physics of the infinite...

AFTERWORD
Theories of Relativity:
What It Means for Philosophy

Charles Alunni

To Mia Zapata and Kristin Hersh

"There is nothing in the whole system of laws of physics that cannot be deduced unambiguously from epistemological considerations."

Sir Arthur Eddington[228]

"The present difficulties of his science force the physicist to come to grips with philosophical problems to a greater degree than was the case with earlier generations."

Albert Einstein[229]

"The reciprocal relationship of epistemology and science is of noteworthy kind. They are dependent upon each other. Epistemology without contact with science becomes an empty scheme. Science without epistemology is – insofar as it is thinkable at all – primitive and muddled."

Albert Einstein[230]

"Relativity, a philosophical and scientific postulate, is a unifying principle as well, a method of constructing the laws of physics, a mode of diagnosing its crises, even a way of thought."

Laurent Nottale[231]

1) Context

Since 1905, and in an always increasing manner, the technical developments of Einsteinian relativity have been accompanied by a *necessary philosophical offshoot* which has taken multiple and varied forms.[232] The great names attached to these theories (special and then general) have from the beginning always strongly insisted on the incalculable consequences that the relativist revolution would have on our ways of thinking, but also, on the fundamental importance of a philosophical tradition entirely renewed by the elaboration of these same scientific ideas.[233] To cite just a few, from the French tradition: Paul Langevin (1872-1946), Jean Becquerel (1878-1953), Charles Nordmann (1881-1940), E. M. Lémeray, Henri Galbrun (1879-1940), Léon Bloch (1876-1947), Élie Cartan (1869-1951), and Georges Darmois (1888-1960); and, elsewhere: Hermann Weyl (1885-1955), Sir Arthur Eddington (1882-1944), Wolfgang Pauli (1900-1958), and Théophile de Donder (1872-1957).[234]

Paul Dirac (1902-1984), author of a wonderful introduction to the mathematics required for understanding general relativity, bears witness to this connection of the scientific with the philosophical:

> The time I am speaking of is the end of the First World War. That war had been long and terrible. . . . Then the end of this war came, rather suddenly and unexpectedly, in November 1918. There was immediately an intense feeling of relaxation. It was something dreadful that was now finished. People wanted to get away from thinking about the awful war that had passed. They wanted something new. *And that is when relativity burst upon us.*
>
> I can't describe it by other words than by saying that it just burst upon us. It was a new idea, a new kind of philosophy, and it aroused interest and excitement in everyone. The newspapers, as well as the magazines, both popular and technical, were continually carrying articles about

it. These articles were mainly written from the "philosophical" point of view. Everything had to be considered relatively to something else. . . . At that time I was sixteen years old and a student of engineering at Bristol University. . . . I was caught up in the excitement of relativity along with my fellow students. We were studying engineering, and all our work was based on Newton. We had absolute faith in Newton, and now we learned that Newton was wrong in some mysterious way. This was a very puzzling situation. Our professors were not able to help us, because no one really had the precise information needed to explain things properly, *except for one man*, Arthur Eddington.[235]

Concerning Eddington, let us recall this typically British anecdote of an absolutely essential author for one conducting a philosophical study of theories of relativity and of attempts to unify these theories.[236] Ludwik Silberstein[237] approached Eddington at the Royal Society's November 6th, 1919 meeting, where he had defended Einstein's relativity with his Brazil-Principe solar eclipse calculations with some degree of skepticism, and ruefully charged Arthur Eddington as one who claimed to be one of three men who actually understood the theory (Silberstein, of course, was including himself and Einstein as the other). When Eddington refrained from replying, he insisted Arthur not be "so shy," whereupon Eddington replied, "Oh, no! I was wondering who the third one might be!"

2) Reception of Theories of Relativity

How does a new theory or discovery get accepted or rejected by a scientific community? Historians of science have published many studies of the reception of Einstein's special and general theories of relativity.[238] For Stephen Brush, there are three kinds of reasons for accepting relativity: 1) empirical predictions and explanations; 2) social-psychological factors; and 3) aesthetic-mathematical factors.

We know that historians and sociologists of science belonging to the analytic school consider that empirical (or "positivist") and social factors are the *only* alternatives for explaining how scientists choose theories. Others, philosophers and, importantly, participants in the scientific enterprise have stressed the importance of the third factor: the fact that a theory must be correct because it is mathematically convincing and elegant, aesthetically pleasing, and expresses a necessary truth about Nature. Albert Einstein expressed this view in his 1933 lecture:

> Nature is the realization of the simplest conceivable mathematical ideas. I am convinced that we can discover, by means of purely mathematical constructions, those concepts and those lawful connections between them, which furnish the key to the understanding of natural phenomena. Experience may suggest the appropriate mathematical concepts, but they most certainly cannot be deduced from it. Experience remains, of course, the sole criterion of physical utility of a mathematical construction. But *the creative principle* resides in mathematics. In a certain sense, therefore, I hold it true that *pure thought* can grasp reality, as the ancients dreamed.[239]

Since Einstein began his 1905 paper with an aesthetic question— the problem of symmetry in Maxwell's theory—it would not be surprising if his followers also gave priority to such issues.[240] The most important advocate of the theory was Max Planck (1858-1947). Planck presented the theory at the physics colloquium in Berlin during the winter semester of 1905-6, and published a paper on it in 1906 (the first publication on relativity other than Einstein's). While Planck did not yet believe that the truth of the theory had been demonstrated experimentally, he considered it such a promising approach that it should be further developed and carefully tested. As a professor at Berlin (at that time one of the major centers of physics) Planck encouraged his students and colleagues to work on relativity theory. As editor of the prestigious journal *Annalen der Physik*, Planck saw to it that any paper on relativity meeting the normal standards would get published.

> According to Goldberg,[241] Planck was attracted to relativity theory because of "his philosophical and ethical convictions about the ultimate laws of reality." He liked the "absolute character with which the physical law was endowed" by relativity theory, such as that the laws of nature are the same for all observers. "For Planck this represented the supreme objectivity toward which science was striving.[242]

Max von Laue (1879-1960), who learned about the theory from Planck, was quickly converted to it and eventually published the first definitive monograph on relativity in 1911: he wrote that relativity "has found an ever-growing amount of attention" despite its inadequate empirical foundation and puzzling assertions about space and time.[243] We can see the radical position of a physicist who gave the first experimental proofs of the general theory: "If there were *no experimental evidence* in support of Einstein's theory, it would nevertheless have made a notable advance by exposing a fallacy underlying the older mode of thought—the fallacy of attributing unquestioningly *a more than local significance* to our terrestrial reckoning of space and time."[244]

One year later, in 1923, Eddington specified:

> The present widespread interest in the theory arose from the verification of certain minute deviations in the theory from Newtonian laws. To those who are still hesitating and reluctant to leave the old faith, these deviations will remain the chief centre of interest: but *for those who have caught the spirit of the new ideas* the observational predictions *form only a minor part of the subject.* It is claimed for the theory that it leads to an understanding of the world of physics clearer and more penetrating than that previously attained.[245]

As he asserted in a famous dictum, one should not "put overmuch confidence in the observational results that are put forward *until they have been confirmed by theory*."[246]

Dirac, who constructed the first successful synthesis of special relativity and quantum mechanics, made it clear that the experimental evidence was not the primary source for his subsequent belief in the theory:

> Suppose a discrepancy had appeared, well confirmed and substantiated, between the [general relativity] theory and observations. . . . Should one then consider the theory to be basically wrong? I would say the answer . . . is emphatically no. The Einstein theory of gravitation has a character of excellence of its own. . . . A theory with the beauty and elegance of Einstein's theory has to be substantially correct. If a discrepancy should appear in some application of the theory, it must be caused by some secondary feature relating to this application which has not been adequately taken into account, and not by a failure of the general principles of the theory.[247]

Einstein himself, though pleased by the eclipse results, gave them little weight as evidence for his theory. According to his student, Ilse Rosenthal-Schneider, after showing her a cable he received from Arthur Eddington about the measurements, Einstein remarked, "But I knew that the theory is correct." When she asked what he would have done if the prediction had not been confirmed, he said, "Then I would have been sorry for the dear Lord—the theory *is* correct." Later he wrote: "I do not by any means find the chief significance of the general theory of relativity in the fact that it has predicted a few minute observable facts, but *rather in the simplicity of its foundation and its logical consistency*."[248]

Many new fundamental and philosophical questions are involved in these principles that will be carried on and extended in the theory of scale relativity of Laurent Nottale.

α) The philosophical essence of theories of relativity

Eddington (as would Hermann Weyl or Gaston Bachelard) summarizes it explicitly:

> It is natural for a scientific man to approach Einstein's theory of Relativity with some suspicion, looking on it as an incongruous mixture of speculative philosophy with legitimate physics. *There is no doubt that it was largely suggested by philosophical considerations*, and it leads to results hitherto regarded as lying in the domain of philosophy and metaphysics.[249]

β) The centrality of the concept of *relation*

> The *relations* unite the *relata; the relata* are the meeting points of the *relations.* The one is unthinkable apart from the other. I do not think that a more general starting-point of structure could be conceived.[250]

γ) The crisis of the concept of *substance*

> In contemplating the starry heavens, the eye can trace patterns of various kinds—triangles, chains of stars, and more figures. In a sense these patterns exist in the sky; but *their recognition is subjective.* So out of the primitive events which make up the external world, an infinite variety of 'patterns' can be formed. There is one type of pattern which for some reason the mind loves to trace wherever it can; where it can trace it, the mind says, 'Here is substance'; where it cannot, it says 'How uninteresting! There is nothing in my line here'. The mind is dealing with a real objective substratum; but *the distinction of substance and emptiness* is the mind's own contribution, depending on the kind of pattern it is interested in recognising. It seems probable that the reason for selecting the

particular type of pattern is that this pattern has (*from its own geometrical character, and independently of the material in which it is traced*) a property known as *Conservation*. Reverting from the four-dimensional world to ordinary space and time, this property appears as *permanence*. *That the mind would necessarily choose for the substance of its world something which is permanent seems natural and inevitable.* The interesting point is that there is no obligation on Nature to provide explicitly anything permanent; *the permanence is introduced by the geometrical quality of the configuration*, which the mind looks out for in whatever Nature provides.[251]

δ) The promotion of the concept of *structure*

The investigation of the external world in physics is *a quest for structure rather than substance*. A *structure* can best be represented as a complex of relations and relata; and in conformity with this we endeavour to reduce the phenomena to their expressions in terms of the relations that we call intervals and the relata that we call events.[252]

Einstein's law is the *simpler* law because it is consistent with what we now know of the general plan of *world-structure*; Newton's law could only be made possible by introducing a novel and specialized feature—*a stratified arrangement of structure*—which is not revealed in any other phenomena.[253]

ε) The idea of a "selective subjectivism" that rejects the idea of a pure subject

The terms space and time have not only a vague descriptive reference to a boundless void and an ever-rolling stream, but denote an exact quantitative system of reckoning distances and time-intervals. Einstein's first great discovery was that there are many such systems of reckoning – many possible frames of space and time – exactly on all fours with one another. No one of these can be distinguished as more fundamental than the rest; no one frame rather than another can be identified as the scaffolding used in the construction of the world. . . . *Nature* offers an infinite choice of frames; *we* select the one in which we and our petty terrestrial concerns take the most distinguished position.[254]

Perhaps a better way of expressing this *selective influence of mind on the laws of Nature* is to say that values are created by the mind. All the "light and shade" in our conception of the world of physics comes in this way from the mind, and cannot be explained without reference to the characteristics of consciousness. . . . The mind has by its selective power fitted the processes of Nature into a frame of law of a pattern largely of its own choosing; and in the discovery of this system of law the mind may be regarded as regaining from Nature that which the mind has put into Nature.[255]

Here, it is essential to establish a parallel with Hermann Weyl's coordinate systems[256] as "the unavoidable residuum of the ego's annihilation."[257] For Weyl, the requirement of a coordinate system needed for the application of analysis to geometry is the residuum of the "pure, sense-giving ego" and its "immediate life of intuition" in the otherwise completely "geometrico-physical" world of relativity theory. It bears the ineradicable trace of transcendental subjectivity that "'constitutes' within itself" the sense of this objective, purely conceptual, world. Of course, according to the principle of general covariance, the choice of coordinate system is essentially arbitrary since the laws of nature are to be formulated in tensor form, valid for all coordinate systems. For Weyl, the choice of a coordinate system, arbitrarily exhibited by an act of the constituting ego, implies as well *a local* choice of unit length or gauge. In physics, more specifically, the purely symbolic world of the tensor fields of classical relativistic physics, is constituted or *constructed* only via subjectivity and *is not understandable as pertaining to objects of completely mind-independent reality,* transcendent to consciousness. The exact determination of the concepts of physics obtained through symbolization cannot be accomplished without the introduction of a coordinate system. It must not be thought, however, that such an "objectification," relative to a coordinate system, is absolute:

> But this objectification [*Objektivierung*] through exclusion of the ego and its immediate life of intuition, is not attained without remainder; the coordinate system, exhibited only through an individual act (and only approximately) remains as the necessary residue of this annihilation of the ego [*das notwendige Residuum dieser Ich Vernichtung*].[258]

In his *Philosophie der Mathematik und Naturwissenschaft* (1926), Weyl returns several times to "the problem of relativity," in one instance repeating the *Ich Vernichtung* passage of *Raum-Zeit-Materie*:

> On the basis of objective geometrical relations, with which the axioms are concerned, it is not possible to determine a point absolutely, but conceptually only relative to a coordinate system, through numbers. For understanding the application of mathematics to reality the distinction between the "giving" [*dem "Geben"*] of an object through individual exhibition on the one side and, on the other in conceptual ways, is fundamental. The objectification through exclusion of the ego and its immediate life of intuition [*Objektivierung durch Ausschaltung des Ich und seines unmittelbaren Lebens der Anschauung*] is not attained without remainder. The coordinate system, exhibited only through an individual act (and only approximately), remains as the necessary residuum of this annihilation of the ego [*das notwendige Residuum dieser Ich-Vernichtung*][259]

ζ) A Relativist Geometry of Space as a Geometrodynamics Project[260]

Take a pair of compasses and twiddle them on a sheet of paper. Is the resulting curve a circle or an ellipse? Copernicus from his standpoint on the sun declares that owing to the FitzGerald contraction the two points drew nearer together when turned in the direction of the Earth's orbital motion; hence the curve is flattened into an ellipse. But here I think Ptolemy has a right to be heard; he points out that from the beginning of geometry circles have always been drawn with compasses in this way, and that when the word 'circle' is mentioned every intelligent person understands that this is the curve meant. *The same pencil line* is in fact a circle in the space of the terrestrial observer and an ellipse in the space of a solar observer. It is at the same time a moving ellipse and a stationary circle. I think that illustrates as well as possible what we mean by *the relativity of space*.[261]

And, more radically:

The difference between space occupied by matter and space which is empty is simply *a difference in its geometry*. There seems to be no reason to postulate that there is an entity of foreign nature present which causes the difference of geometry; and if we did postulate such an entity it would scarcely be proper to regard it as physical matter; *because it is not the foreign entity but the difference of geometry which is the subject of physical experiment*.[262]

η) The External World of Physics as "the viewpoint of no one in particular"

Eddington was also heretical enough to accept Weyl's generalization of Einstein's theory and to generalize it further, for epistemological reasons essentially similar to Weyl's. Eddington sought the same goal of constituting the "real world of physics" by reconstructing relativity theory within a differential geometry capable of yielding only objects that are a *synthesis of all aspects*" present to *all* conceivable observers. The external world of physics might be defined in this way as a world conceived *"from the viewpoint of no one in particular,"* a standpoint both necessary and sufficient for objective representation in physics. The epistemological significance of relativity theory lay in showing that the attempt to portray the physical world from this *impersonal* perspective resulted in its *geometrization*. In turn, the physical knowledge captured in such a portrayal is knowledge only of that world's structure. Physics could be about no other world than that *expressly incorporating all viewpoints at once*, an "absolute world" as opposed to the "relative" world of each individual perspective, that is, any "conceivable observer." The relation between the *relative* and the *absolute* is mathematically captured by the tensor calculus and physical knowledge accordingly must be represented in the form of tensor identities through a method Eddington called *"world building."*

There are two important aspects to this, which relate to the *structuralist* and *subjectivist* components of Eddington's thought, respectively. By matter as the putative cause of irregularities in the field, Eddington meant *matter as substance*, and thus this construction was seen as eliminating substance from our ontology in favor of relational structures. Secondly, "matter," in this new sense, became dependent on the mind, since "Matter is but one of a thousand relations between the constituents of the World, and it will be our task to show why one particular relation has a special value for the mind."[263]

The point of view of the observer is declared by Eddington "the parochial standpoint": "We must try another plan. I do not think we can ever eliminate altogether *the human element* in our conception of nature; but we can *eliminate a particular human element*, namely, this framework of space and time."[264] Using the example of a chair, considered as a solid object which cannot be identified "with any one of our two-dimensional pictures of it, *but giving rise to them all* as the position of the observer is varied," Eddington specifies:

By sheer necessity our brains have been forced to construct the conception of the solid chair *to combine these changing appearances*. But we do not vary our motion to any appreciable extent and our brains have not hitherto been called upon to combine the appearances for different motions; thus the effort which we now ask the brain to make is a novel one. That explains *why the result seems to transcend our ordinary mode of thought.*[265]

Let us finally note that these questions echo the remarkable *Eddingtonian conception of tensors* developed in his *Tarner Lectures*:

Since physical knowledge must in all cases be an assertion of the results of observation (actual or hypothetical), we cannot avoid setting up a dummy observer; and the observations which he is supposed to make are subjectively affected by his position, velocity and acceleration. The nearest we can get to a non-subjective, but nevertheless observational, view is to have before us the reports of all possible dummy observers, and pass in our minds so rapidly from one to another that we identify ourselves, as it were, with all the dummy observers at once. To achieve this we seem to need a revolving brain.

Nature not having endowed us with revolving brains, we appeal to the mathematician to help us. He has invented a transformation process which enables us to pass very quickly from one dummy observer's account to another's. The knowledge is expressed in terms of tensors which have a fixed system of interlocking assigned to them; so that when one tensor is altered all the other tensors are altered, each in a determinate way. By assigning each physical quantity to an appropriate class of tensor, we can arrange that, when one quantity is changed to correspond to the change from dummy observer A to dummy observer B, all the other quantities change automatically and correctly. We have only to let one item of knowledge run through its changes – to turn one handle – to get in succession the complete observational knowledge of the dummy observers.

The mathematician goes one step farther; he eliminates the turning of the handle. He conceives a tensor symbol as containing in itself all its possible changes; so that when he looks at a tensor equation, he sees all its terms changing in synchronised rotation. This is nothing out of the way for a mathematician; his symbols commonly stand for unknown quantities, and functions of unknown quantities; they are everything at once until he chooses to specify the unknown quantity. And so he writes down the expressions which are symbolically the knowledge of all dummy observers at once – until he chooses to specify a particular dummy observer.

But, after all, this device is only a translation into symbolism of what we have called a revolving brain. A tensor may be said to symbolise absolute knowledge; but that is because it stands for the subjective knowledge of all possible subjects at once.[266]

We cannot develop this matter here, but it is interesting to note that these views are connected with the concepts of *shadow* and *skeleton* held by both Einstein and Bachelard:

If you want to fill a vessel with anything you must make it hollow. . . . Any of the young theoretical physicists of today will tell you that what he is dragging to light as the basis of all the phenomena that come within his province is a scheme of symbols connected by mathematical equations. . . . Now a skeleton scheme of symbols is hollow enough to hold anything. It can be—nay it cries out to be—filled with something to transform it from skeleton into being, from shadow into actuality, from symbols into the interpretation of the symbols.[267]

It is this metaphor of the *skeleton* that Einstein would use, for example, to demonstrate that with his theory, gravity moves from the periphery to the true center of physics: "The gravitational field . . . appears, as it were, *as the skeleton* from which everything hangs."[268]

3) The Theory of Scale Relativity (TSR) of Laurent Nottale

On this basis of the different philosophical postulates, essential to the existence of the special and general theories of relativity, the theory of scale relativity sets forth, pursuing and continuing, or rather rectifying the initial work of Albert Einstein and his disciples.

A few words on this point.

For Laurent Nottale, the principle of scale relativity *generalizes* the claim of Einstein: the same laws of nature apply whatever state of motion an object is in, *but also whatever the scale of the coordinate system may be.* Spatiotemporal resolutions possess the same relativity as motion: just as one would not be able to define an interval of length or of time in an absolute manner, only *a relation* between two scales has meaning. When one applies to spacetime itself this idea that physical quantities explicitly depend on this resolution, one is led to the geometric concept of *fractals.* In the beginning of the 1980s, Garnet Ord, of the University of Ontario, and Laurent Nottale have independently proposed that *quantum properties* arise from the fractal nature of microscopic spacetime.

But what about the corresponding mathematics? Since Isaac Newton, we have used the *differential method* to put physical phenomena into equations: we decompose a *complex object* into its simpler parts. This simplicity allows a *local* description, a differential, which, after integration, provides the *global* properties of this object.

However, this *method* is no longer valid when the parts, *instead of becoming more simple*, become *more complex* or different from the object from which one began. This is exactly what happens in particle physics: when we observe an object with a particle accelerator, which replaces the microscope at these scales, *new structures* appear whenever there is an enlargement. *Quantum mechanics describes this behavior.* Einstein's principle of "general" relativity, based on differentiability, is then *necessarily* incapable of taking into consideration quantum effects, which are based on *nondifferentiability.*

α) Relativity of coordinate systems

We know that the theory of special relativity is the *general solution* to the relativistic problem of inertial motion, which one could already ask in Galileo's age: what are the *laws of transformation* of the inertial coordinate systems (moving with a *constant* relative velocity with respect to each other) that satisfy the principle of relativity? These are the Lorentz transformations. One of the improvements that Laurent Nottale adds here to classical theory is that to obtain these laws, one does not need to add to the principle of Galilean relativity the postulate of the *invariance of the speed of light* in a vacuum, as Einstein did in 1905: in the establishment of their *general* form, a constant c appears, which one can then identify as the speed of *any massless particle in a vacuum.* The Galileo transformation is then no longer anything but a special case of the Lorentz transformation, which corresponds to the choice of an *infinite constant c.* The principle of relativity entails the Lorentz transformations, and thus the concept of spacetime.

β) Deducing gravity from relativity

From the perspective of Einstein's *general* theory of relativity, even the existence of a gravitational field is no longer absolute, but dependent on the motion of the coordinate system under consideration: in a system in free fall within this field, gravity completely disappears.[269] This is what astronauts experience as "weightlessness." Gravity can be understood as the set of manifestations of curvature, necessitating the transition from a *flat*, Euclidean spacetime, to a *curved* spacetime. In general relativity, the motion of a particle under the influence of an arbitrarily complicated gravitational field is described by a *locally*

290

inertial motion (at constant velocity) along the geodesics of a *curved* spacetime. The Einstein equations which connect the curvature of spacetime with the distribution of energy and matter are *the simplest and most general equations* which are *invariant* under *continuous and twice-differentiable* transformations of coordinate systems: thus, Einsteinian relativity *dictates* the existence of gravity as well as *the form* of the equations which describe it.

γ) The axioms of quantum mechanics

Quantum theory relies on *axioms*, deduced from microphysical experiments which are impossible to explain using classical concepts. The paths of particles are not observable: *the paths are absent*. Quantum theory combines the three elements of "probability," "wave," and "particle" into *one sole theoretical object*, the wave function. Erwin Schrödinger and Werner Heisenberg wrote the equations which govern it. However, these equations and the "correspondence principle," which associates operators acting on a wave function to observable magnitudes, *cannot be deduced starting from a first principle*, but are supposed *a priori*.

In current quantum theory, *spacetime is flat*, as in special relativity. Nevertheless, the development of ideas in physics (Leibniz, Mach, and Einstein) has deprived all scientific meaning from the idea of a space independent of its contents. Is it not a contradiction to allow for, on the one hand, the existence of microphysical objects with *universal* quantum, nonclassical properties, and on the other hand, to consider that the *framework which contains these objects* is in no way modified?

δ) Nondifferentiable spacetime

In Richard Feynman's 1965 book written with Albert Hibbs, Feynman describes the virtual paths typical of a quantum particle:

> The important paths for a quantum-mechanical particle are
> not those which have a definite slope (or velocity) everywhere,

but are instead quite irregular on a very fine scale. . . . Typical paths of a quantum-mechanical particle are highly irregular on a fine scale. . . . Thus, although a mean velocity can be defined, no mean-square velocity exists at any point. In other words, the paths are nondifferentiable.[270]

This introduction of nondifferentiability into physics is all the more remarkable considering that, during the same period, Einstein had himself explicitly imagined that a realistic approach to the quantum problem could follow the same route. In 1948, he wrote in a letter to Wolfgang Pauli:

This complete description would not be limited to the fundamental concepts used in point mechanics. I have told you more than once that I am a fierce partisan not of differential equations, but of the principle of general relativity, whose heuristic force is indispensable to us. Yet, in spite of much research, I have not succeeded at satisfying the principle of general relativity otherwise than using differential equations; perhaps someone will discover another possibility, if they look with enough perseverance.[271]

Laurent Nottale is precisely this "someone"!

From a philosophical point of view, it is important to note here that as early as 1940, in a chapter titled "Elementary spatial connections. Non-analyticity," Bachelard foretold the *fractal* character of his epistemology by drawing the attention of the philosophic community to what would much later become the concept of "quantum scaling."

Mechanics has slowly freed itself from the concept of a *jet* or a throw, it has not sufficiently thought about the circumstances of a *traject* or course. Now the trajectory of a micro-object is a closely circumscribed *traject*. Continuity of the whole must not be postulated; the connection must be examined link by link.[272]

This is the hypothesis of *non-analyticity*, first postulated in his thesis of 1927, and which he would detect even in the foundations of Wolfgang Pauli's *exclusion principle*.[273] The second step would be for him to extend wave mechanics:

> As soon as one abandons the very special claim of mathematics to analyticity, as soon as one accepts the non-analytical constitution of trajectories, it becomes clear that one can set up connections which, in spite of their artificial character, permit information about certain properties of trajectories in wave mechanics.[274]

He took his example of "non-analytic paths" from "the simple yet profound work of Adolphe Buhl."[275] Working independently of Paul Dirac, whose work on the *Zitterbewegung* ["trembling motion"] of the electron whose fluctuation makes its physical trajectory oscillate to such an extent that it has an "average velocity" as low as experiment measures it, Adolphe Buhl raised the fact starting in 1934 that such a trembling is already imaginable in classical mechanics, and that if we were to stop seeing only "taut" paths, *we would discover an infinite number of random trajectories that corresponded well to the equations of mechanics.* Bachelard recognized in this, through a detailed technical analysis, an applied rationalization of the Heisenberg principle.[276] But it is by a true return to the consequences (for philosophical thought) of this Buhlian hetero-induction that he would *explicitly*, and definitively, lay the groundwork for a *fractal axiomatic*, a true surrational manifesto:

> The ingeniousness of Buhl's memoir lies in the fact that he really integrates the ambiguity *all the way along* the integral curve whereas intuition lazily confines itself to attaching it to the origin of the trajectories. Let us, therefore, take stock of the freedom we have. . . . We see that *a saw-toothed path*

appears, each of the teeth representing a little arc *which responds to the obligations of the problem*. The number of teeth can, moreover, be increased at will, since the track segments are as small as one cares to make them.

Moreover, this trajectory, made up of vibrations, retains some important properties. It keeps continuity, and it keeps the length of the trajectory which common intuition would have selected, since all its fragments satisfy the isometric condition. But in spite of the continuity, the infinitely small appears as infinitely broken up, intimately ruptured, without any quality, without any concern, without any destiny passing from one point to the neighboring point.[277] It seems as if, along a Buhlian trajectory, motive power had nothing to transmit. It is really the most gratuitous movement. But along a trajectory of natural intuition, on the contrary, the motive power transmits what it does not possess; it transmits the cause of its direction, a sort of coefficient of curvature which makes it impossible for the trajectory to change sharply. . . . The objection will be raised that common experience affords no examples of hesitant trajectories. We shall even be accused of a veritable initial contradiction since we adopt a non-analytical solution for a problem posed within the framework of analytical data. Let us look more closely at these two objections.

To be sure, everyday experience only yields analytical trajectories and we can actually only draw analytical curves. But the argument comes back on itself. As Buhl justly observes one can perfectly well inscribe a sub-design within the thickness of an experimental line itself, a wavering line, a veritable arabesque which specifically represents indeterminacy of the second degree of approximation. In short *any linear structure whether real or realized encloses finer structures*. This refinement is without limit. It is really a matter of "an indefinitely fine structure." And so we see, within the domain of pure geometry, that concept of *fineness of structure* which has played so important a role in the progress of spectrography. *And this is no mere metaphorical juxtaposition.*[278] It really appears that the works of Buhl illuminate *a priori* a great many problems of micromechanics and of microphysics. In these fine structures there appear, be it said in passing, the famous *continuous functions without derivatives*, continuous curves without tangents. They are the mark of the perpetual indeterminacy of the trajectory of fine structure. . . .

But we also have to counter the accusation that we have been inherently contradictory. Is not the genesis of isometric trajectories actually a basic differential equation? Does one not, therefore, posit the existence of a derivative at all the points of the integral curve? How then can one offer a continuous curve, but without derivative, as the solution of an equation which is involved with the elementary intuition of the derivative?

This second objection, like the first, must be turned back against the partisans of natural intuitions. . . . Here, as Buhl points out, the methodological contradiction is, all things considered, nothing but the result of an unjustified claim inherent in the postulates of the research. . . . Naturally, if the proposed problem admits solution by a saw-toothed trajectory it also admits (subject to some modifications suggested by Buhl) a return of the trajectory upon itself, a folding back. All this is by way of demonstration that the conditions which establish the path of a mobile point can be endlessly diversified even when restrained by a law as simple as that of an isometric trajectory, and that irreversibility in particular is a very special notion which loses the greater part of its normal meaning when it comes to a study of second degree approximation. This is a conclusion to which microphysics has become accustomed.[279]

Then Bachelard, through a movement of recursion or of inductive retro-construction, moves on to the *quantum* domain:

> Thus it is very interesting to observe, with Adolphe Buhl, that the complementary uncertainties organized by Heisenberg find a useful illustration in Buhlian propagation. Actually the theme of the principle of Heisenberg can be connected to the entirely geometric fine intuitions organized by Buhl without the necessity of appending dynamic circumstances. . . . In the problem of Buhlian "ray," at the level of indefinitely fine structure a *precise* concept of a tangent at a *precise* point makes no sense. It is impossible to attach a tangent to a well-defined point. And vice-versa, if one gives a well-defined tangential direction one cannot discover the point which will receive it. . . . In a humorous vein one might say that the tangent is flighty and space is dotty in every sense of the words. The two kinds of madness are correlative. There is an opposition between precision of position and precision of direction.
>
> The Buhlian trajectory is thus enriched by its value as a supplementary diagram. . . . The work of Adolphe Buhl thus brings about a veritable rationalization of the Heisenberg principle.[280]

The problem posed by such a change in the geometric paradigm seems nevertheless to be extremely difficult: would the abandonment of differentiability not require one to abandon differential equations, an essential methodology for all physics? Fortunately, another way, that followed and developed by Laurent Nottale, is possible which, in a stunning manner, is brought back to the preceding, all while providing a mathematical tool that allows one to *describe nondifferentiability with the assistance of differential equations.*

The key for the solution is found by *interpreting Feynman's work in terms of fractals.* Without entering into the details, this results from a theorem of the mathematician Henri Lebesgue: a curve of finite length is differentiable almost everywhere. Inversely, if a continuous curve is nondifferentiable almost everywhere, it is necessarily of *an infinite length*. Abandoning the arbitrary hypothesis that a curve in spacetime is differentiable, while maintaining its continuity, implies an explicit dependence on resolutions.[281] *Relativity extended to nondifferentiable motion is thus equivalent to scale relativity.* Nottale simply requires that the equations written in such a nondifferentiable spacetime satisfy the *covariance principle*, meaning that they keep *the same form* as in the differentiable case.

One initial way of discovering the form of these scale laws is to *postulate* that they are the *simplest possible*.[282] One writes a first-order differential equation over an infinitesimal change in resolution: its solution is the length of a fractal curve of constant dimension. Thus, the fractal functions of constant dimension, which diverge in a power relation as a function of the resolution, are the simplest forms of laws which depend explicitly on the scale. *This is precisely the behavior obtained by Feynman for quantum paths.*

ε) Deducing quantum physics from relativity

Laurent Nottale then *deduces* the principal axioms of quantum mechanics from the fractal spacetime concept, since *nondifferentiability requires the probabilistic character of the description.* In Einstein's theory, the path of a free particle is a geodesic in spacetime. It would be the same in a fractal spacetime. Nevertheless, the presence of fluctuations at small scales makes *the number of geodesics infinite*, all of which, *by definition*, are equally probable: the only possible prediction concerning the motion of the "particle" should be deduced from the density of *this infinite family of* geodesics (a density which is variable and generates a probability density).

Such a statement is incomplete, since *the fractal approach also transforms the concept of the elementary particle.* In current quantum theory, the electron, from the perspective of its particle-like nature is a point. In fractal spacetime, one gives up the idea of points with mass and considers "particles," with the dual nature of wave and point, as the set of properties of geodesics. Here we arrive at a connection with a true *geometrodynamics* (of the Wheeler-Eddington type): the "internal" properties finally resulting from symmetries connected to scale transformations take on a *geometric significance*, in the sense of nondifferentiable geometry. The concept of a particle no longer concerns an object which "has" a spin, a mass, or a charge, but is reduced to the *geometric structures* of the fractal geodesics in nondifferentiable spacetime.

Two examples:

— the indistinguishability of identical particles is an immediate consequence of their identical fractal paths. These paths possess no proper characteristic which would allow us to distinguish them. A set of many particles is no longer considered as a collection of individual objects in the classical sense: it is a *new* object, a *network* of geodesics which *possesses its own geometric properties*.[283]

— the meaning of wave-particle duality is different from the usual interpretation in quantum theory. In classical quantum theory, the wave function describes a wave-particle. Here, its particle-like nature is a consequence of quantization and of elementarity (in agreement with quantum physics where particles are quanta). If a photon reaches a screen, it cannot be divided and can only be absorbed by a single electron of a single atom. Here is the cause of the single "click" which leads to the concept of the particle. But the geodesics which arrive at this point (at a given resolution) remain at an infinite number. This is clear for example in the Young double-slit experiment conducted one particle at a time.[284]

Moreover, the sheaf of *possible* geodesics, the only methodology which enables us to make predictions, conveys the *wave-like properties*. In conclusion, not only is the TSR, anchored on extremely profound philosophical principles, an *extension* of Einstein's special and general relativity, by the addition of a supplementary dimension, but it also opens the field of unification with quantum theory, which, in this framework, is all *naturally* deduced (the Schrödinger equation, the Dirac equation . . . within a fractal spacetime): we are very far from the extremely complex contrivances and purely mathematical conjectures such as string theory and superstring theory, conjectures which address a very different problem, since they base themselves within the framework of quantum mechanics without seeking to understand or provide a foundation for it.

The production of theoretical texts and experimental results covering not only the domain of physics, but including a large number of related disciplines such as biology and geology is very impressive. The situation is clear regarding the present and future power of this theory, and yet...

4) The Question of the "Reception" of the TSR

In conclusion, instead of elaborating a sociological theory of the reception of this theory, one should simply highlight the situation which leaves the philosopher a prohibited figure, and which should, at the very least, attract the attention of the physicist.

During the nineties, and up until 2005, the TSR benefited from great interest marked by different publications seeking to introduce the theory to the public, to spread awareness of it and to popularize it. One can cite issue number 936 of *Science & Vie* ["Science and Life"], from September 1995, for which Laurent Nottale appears on the cover: "Fifty years after Einstein a scholar sheds light on the mysteries of the universe." At the same time, also in September 1995, the French edition of *Scientific American* devoted a significant section to *scale relativity*, with an article by its creator entitled "Fractal spacetime; The theory of scale relativity generalizes Einstein's principle of relativity; Quantum mechanics becomes a consequence of relativity."[285]

Another issue, this time dedicated to alternatives to the standard model of physics, appeared in 2005, the year of Einstein. Again Laurent Nottale was on the cover of *Science & Vie* alongside Alain Connes, Carlo Rovelli, and Gabriele Veneziano in a group photo, under the title "Relativity is a hundred years old; Moving beyond Einstein; They want to reinvent space and time."[286] Inside, the magazine simultaneously presented approaches in noncommutative geometry, loop quantum gravity, string theory, and scale relativity.

After these signs of public interest in TSR, not only was there a sudden halt in coverage, but there was a series of attacks by the physico-mathematical community against the theory . . . Thus, I will pose a multi-faceted question to this community:

α) Why has interest in this theory been of concern for philosophers foremost (but not only)?[287] Part of the answer can be found in the first part of this text;

β) Why was the purely objective, meaning scientific and rational debate concerning the TSR silenced in a Stalinesque trial supported by assorted malodorous rumors about its author?

γ) Why have the same people, initially welcoming of the theory, silently distanced themselves, then actively opposed it as soon as experimental evidence materialized?

Would the very French syndrome of the "*grandes écoles*" [great schools] be one of the primary reasons: the *École Centrale de Paris*, from which Laurent Nottale came, pitted against the symbolic power of the *École Normale Supérieure*, the nucleus of the contempt and the attacks?

δ) Why the *ad hominem* attacks and threats against those who were closely interested in this theory?

ε) Why the shy silence on behalf of the less hostile?

ζ) And finally, an essential question, what has happened to this community of physicists so that it only puts forth such arguments of authority as answers of the kind: "No, I haven't studied this theory, but I've heard people say that..." in response to rational debate concerning the "falseness," the "faults," or the "errors of calculation" for which nobody has ever provided the least amount of evidence?

Here is not the place to answer these questions which concern the community and which are nothing but the symptom of a true crisis in physics, tied to the loss of historical memory of the discipline, to the forced march toward globalization (collusion of committees of peer reviewers), to the most savage competition for research funding, the massive financial stakes of contemporary experimental physics, before the serious alternative proposed by the TSR, which has the potential to open phenomenal possibilities of scale economies and new developments in research. And this is to say nothing of its results on the terrain of a number of other disciplines: economics, medicine, biology, paleontology, geography, not to mention chaos theory in physics, turbulence, and complex systems.[288]

Therefore, just one more word: *read and study* this marvelous theory, allow yourself to be moved by its beauty, its depth, and its major experimental implications, which are nothing less than *fundamental for the future of science and for philosophy.*

Notes

[1] Galileo Galilei, *Dialogue Concerning the Two Chief World Systems, Ptolemaic and Copernican*, Trans. Stillman Drake (Berkeley: University of California Press, 1967, 2nd Edition.) The First Day, 9.

[2] Nicolaus Copernicus, *On the Revolutions of the Heavenly Spheres*, Trans. A. M. Duncan. (New York: Barnes and Noble, 1976) Book One, 44.

[3] Galilei 116.

[4] Galilei 186-188.

[5] Galilei 171-172.

[6] Newton was trained particularly through his study of the *Principles of Philosophy* and the *Dialogue Concerning the Two Chief World-Systems*. Cf. Colin Ronan, *Histoire Mondiale des Sciences*, (Paris: Seuil, 1988).

[7] The theory of scale relativity suggests an answer to this problem: in short, the masses which curve spacetime are those of elementary particles, which are nothing but spacetime itself in this theory (namely, they are identified with the geodesics of a fractal spacetime). Then the fractality of spacetime turns to curvature at large scales.

[8] Leibniz's view of differentials has known a new life with Robinson's mathematical construction of "Non Standard Analysis," in which one can define infinitesimal numbers (which are smaller than any finite number)— and also infinite numbers.

[9] Cf. Christiane Vilain, "Huygens and Relative Motion," in *Relativity in General*, Eds. J. Diaz Alonso and M. Lorente Paramo (Éditions Frontières, 1994).

[10] Ronan 583.

[11] Indeed, in a given frame of reference including space and time, rest corresponds to no change in space while time does change; in motion, both space and time change; therefore, the transformation from the rest *line of universe* to the motion amounts to a rotation in spacetime.

[12] Jules Leveugle, *La Relativité, Poincaré et Einstein, Planck, Hilbert* (Paris: Editions L'Harmattan, 2004). 30.

[13] Henri Poincaré, *Science and Hypothesis* (London: Walter Scott, 1905) 90, 116-17, 211-12.

[14] Poincaré's Note, in the June 5, 1905 issue of the *Comptes Rendus de l'Académie de Sciences*, 25-26.

[15] Poincaré, *Comptes rendus* 25-26.

[16] Poincaré, *Comptes rendus* 26.

[17] Poincaré, *Comptes rendus* 27.

[18] Poincaré, *Comptes rendus* 27.

[19] Poincaré, *Comptes rendus* 34.

[20] Poincaré, *Comptes rendus* 29.

[21] Quoted in Abraham Pais, *Subtle is the Lord: The Science and the Life of Albert Einstein* (New York: Oxford University Press, USA, 2005). 100.

[22] Poincaré, *Comptes rendus* 30.

[23] Poincaré, *Comptes rendus* 30-31.

[24] Poincaré, *Comptes rendus* 38.

[25] Poincaré, *Comptes rendus* 34.

[26] Poincaré, *Comptes rendus* 33.

[27] Henri Poincaré "The Theory of Lorentz and the Principle of Reaction," *Archives néerlandaises des Sciences exactes et naturelles* ser. 2, v. 5, 252-278 (1900). Trans. Steve Lawrence. http://physicsinsights.org/poincare-1900.pdf. 8.

[28] Poincaré, *Comptes rendus* 37.

[29] Poincaré, *Comptes rendus* 38.

[30] In modern notation, the (wrong) transformation obtained by Lorentz is written:

$x' = x/\sqrt{(1-v^2/c^2)}$, $y' = y$, $z' = z$, $t' = t \sqrt{(1-v^2/c^2)} - (vx/c^2)/\sqrt{(1-v^2/c^2)})$, which is not the transformation of special relativity.

[31] $x' = (x - v\, t)/\sqrt{(1-v^2/c^2)})$, $y' = y$, $z' = z$, $t' = (t - vx/c^2)/\sqrt{(1-v^2/c^2)}$.

[32] That is, a "group" in the mathematical sense of group theory, which has played a greater and greater role in modern physics.

[33] It seems that Einstein was aware of the work published by Poincaré up to 1902, but not afterward.

[34] Albert Einstein, *Albert Einstein: Autobiographical Notes*, Trans. Paul Arthur Schilpp, Ed. La Salle, (Chicago: Open Court, 1979) 49-51.

[35] Albert Einstein, "How I Created the Theory of Relativity," *Physics Today* (Aug. 1982, Trans. Yoshimasa A. Ono) 45-47, 46.

[36] Albert Einstein, "On the Electrodynamics of moving bodies," *The Principle of Relativity: Original Papers by A. Einstein and H. Minkowski*. Trans. M. N. Saha and S. N. Bose (Calcutta UP, 1920) 1-34, 2.

[37] "On the Electrodynamics of moving bodies" 2.

[38] It is the very *principle* of relativity of Galileo that is supposed to be valid here, not the Galilean mathematical laws of relativity, which have to be changed.

[39] "On the Electrodynamics of moving bodies" 2.

[40] In fact, the Galilean transformation satisfies it as well, but it corresponds to the special case of an infinite speed of light.

[41] The Lorentz factor is written: $\gamma = 1/\sqrt{(1-v^2/c^2)}$.

[42] The Minkowski invariant is written $ds^2 = c^2dt^2 - (dx^2 + dy^2 + dz^2)$. The quantity ds/c gives the proper time, that is, the time that passes for a clock contained within the system under consideration.

[43] Kinetic energy is written in its relativist form: $E = mc^2/\sqrt{(1 - v^2/c^2)}$.

[44] L. Nottale, "The theory of scale relativity," *International Journal of Modern Physics*, A 7, 4899-4936 (1992); *Fractal Space-Time and Microphysics*, (Singapore: World Scientific 1993), Chapt. 6.

[45] The Galilean laws are a degenerate version of those of Einstein-Poincaré, for they correspond to an infinite speed c.

[46] Mathematically, it should be an *internal* law of composition, which is one of the axioms of the structure of a group.

[47] See Nottale, "The Theory of Scale Relativity."

[48] Since 1985, it has been set exactly at 299,792,458 m/s.

[49] A nanosecond corresponds to about 30 cm.

[50] Einstein, "How I Created the Theory of Relativity," 47.

[51] Unless we wait several hours, which would allow us to verify that we are not really in an inertial system, but in a rotating system with the Earth. The inertial forces that result exist, but are too weak to be felt. Similarly, over several months, it would be the movement of the Earth rotating around the Sun which would manifest, etc.

[52] The arc contained within a plane passing through the two points to be connected and the center of the sphere.

[53] At a close factor of proportionality which boils down to Newton's gravitational constant.

[54] Here we ignore the right angles so as to concentrate on the properties of the square similar to those of the circle.

[55] The direct observation of gravitational waves since 2015 in the LIGO and Virgo detectors has brought a new and spectacular proof of Einstein's theory of gravitation.

[56] The English translation reads: "In general, Laws of Nature are expressed by means of equations which are valid for all coordinate systems, that is, which are covariant for all possible transformations" (Einstein, "Foundation of the Generalized Theory of Relativity").

[57] Cf. Steven Weinberg, *Dreams of a final theory: the search for the fundamental laws of nature* (London: Vintage,1993) and Stephen Hawking, *A Brief History of Time* (New York: Bantam, 1998).

[58] To illustrate this point of view, we can recall that as the equations of physics become more fundamental, they become more difficult to solve. Thus, the two-body problem of gravity (that of the motion of a binary star) is simple in Newtonian theory, but unsolvable in an exact manner in Einstein's theory. One might imagine that if one day the equations of a totally unified field are written, even the one-body problem will no longer have an exact solution!

[59] The three laws are: 1) Planets travel in ellipses of which the Sun is one of the foci; 2) the vector radius that is drawn between the Sun and a planet covers equal areas in equal times; 3) the relation of the cube of the semimajor axis over the square of the period is the same for all planets in the solar system.

[60] The emergence of the laws of quantum mechanics from a nondifferentiable continuous (then fractal) spacetime in the theory of scale relativity comes, at some level, under such a statement.

[61] Here the word "absolute" is not used in its sense of "unconditioned" or "not-relative," which would just be contradictory with the viewpoint of relativity. It rather refers to looking for what does not change within change.

[62] Invariants are of two kinds: some are conservative quantities which do not change during the evolution of a system (such as energy, momentum, or angular momentum), but depend on the coordinate system; some other invariants are quantities which do not depend on the choice of the coordinate system. For example, the length of a ruler does not depend on its orientation in space, while its projections on the three axes, horizontal and vertical, do depend on it.

[63] Galilei 116.

[64] Albert Einstein, *Oeuvres choisies, tome 1: Quanta* (Paris: Seuil/CNRS, 1989) 249.

[65] Let us note that, as Weinberg has analyzed in depth, it is not a case of writing arbitrarily complicated equations which contain in themselves all possible effects issued from the choice of a coordinate system, itself arbitrarily complex. Rather, the equations must keep, within whatever reference system, the form that they had in the simplest system. In this form, the principle of covariance boils down to the equivalence principle which Einstein had used to construct his theory of gravitation. In fact, the simplest laws are those of free inertial movement; yet the (local) equivalence principle of a gravitational field or a field of acceleration leads precisely to the statement according to which particles in free fall in a gravitational field are in fact only submitted to inertial laws (see Part One). One can also distinguish a "weak covariance," within which fundamental equations, in a more complex situation, keep the form

they had in the previous, simpler situation; and a "strong covariance," where the form of equations would be those of the simplest situation possible, inertial motion in a vacuum.

[66] This fact is expressed mathematically by the Pythagorean relation: whatever the values measured are for the dimensions of the ruler projected along three axes perpendicular between them, x, y, and z, the square of the length is invariant (meaning it is independent of the system of axes) and given by the sum of the squares of each coordinate, which is expressed mathematically by the relation $l^2 = x^2 + y^2 + z^2$. None of the individual coordinates x, y, and z are invariant; together, they form a vector quantity (x, y, z) of which they are the components. This vector is defined independently of the reference point, even though its manifestation within a reference point makes projected components appear of which the numerical values change from one coordinate system to another.

[67] Vilain, "Huygens and relative motion."

[68] Gottfried Wilhelm Leibniz *The philosophical works of Leibnitz : comprising the Monadology, New system of nature, Principles of nature and of grace, Letters to Clarke, Refutation of Spinoza, and his other important philosophical opuscules, together with the Abridgment of the Theodicy and extracts from the New essays on human understanding : translated from the original Latin and French* Trans. George Martin Duncan (New Haven: Tuttle, Morehouse, and Taylor, 1890) 213, original emphasis.

[69] However, the account of the relativity of all things implies a reformulation of the question. It is not true that "there is something rather than nothing" since, in the absolute, there is absolutely nothing. The question should therefore be written *"Why is there something relative rather than nothing?"*

[70] One might object that the equations of Einstein in a vacuum (meaning without source) allow as a trivial solution the spacetime of Minkowski, which is thus flat, absolute, and empty. But even the emptiness of this space excludes the presence of an observer. The sole fact of including the observer in the description necessitates the existence of a curvature for spacetime. Minkowski's spacetime thus finds itself excluded from acceptable global solutions (all in describing spacetime locally, within a frame of reference in free fall).

[71] The difficulty of the task has increased again with the discovery of the strong and weak nuclear fields.

[72] We must also refer to non-commutative geometries, developed in particular by Alain Connes.

[73] Erwin Schrödinger, "Quantisation as a Problem of Proper Values, Part II," *Collected Papers on Wave Mechanics*, Second Edition (New York: Chelsea Publishing, 1978) 13-40, 13.

[74] According to which the square of its modulus gives the probability of the particle's presence.

[75] Richard Feynman, *QED: The Strange Theory of Light and Matter* (Princeton: Princeton University Press, 1985) 5, 9, 10.

[76] This relation is made with the help of the mathematical transformation called the Fourier transform.

[77] Or alternatively, half of the reduced Planck constant $\hbar = 1.055 \times 10^{-34}$ J sec.

[78] G. Lochak, S. Diner and D. Fargue, *L'objet Quantique* (Paris: Flammarion, Nouvelle Bibliothèque Scientifique, 1989).

[79] A body's angular momentum is of the same order of magnitude as the product of its mass, the square of its radius, and the angular speed of rotation. If the radius is zero, the angular momentum would thus seem to then be zero as well.

[80] In units of $\hbar$: this is made possible because quantum mechanics predicts that the differences of angular momentum must be integer multiples of $\hbar$, which can arise from values like $0, \pm1, \pm2\ldots$, but also $\pm1/2, \pm3/2$, etc.

[81] Note, nevertheless, that Louis de Broglie had obtained these results in his PhD thesis more than a year before Bose and Einstein. See de Broglie and Lochak.

[82] See Victor Weisskopf, *La révolution des quanta*, (Paris: Hachette, 1989).

[83] Its mass is 9.109×10^{-28} g, its charge 1.602×10^{-19} Coulomb, and its spin ½ in units of $\hbar$. The square of the charge, expressed in quantum units, is a dimensionless constant called the fine structure constant, $\alpha = e^2/\hbar c$, which

has an inverse value of 137.0359997 ± 0.0000001. The origin of this pure fundamental number is one of the main mysteries of physics.

[84] This has also been experimentally verified: everything indicates that the collapse of the wave function is indeed instantaneous for this experiment, in agreement with the pre-existence before the measurement of the cancelling out of the sum of spins.

[85] See Pais, *Subtle is the Lord: the Science and the Life of Albert Einstein* and R. Deltete, R. Guy, "Einstein's Opposition to the Quantum Theory," *American Journal of Physics* (1990) 58, 673.

[86] A. Einstein, "On the Method of Theoretical Physics," *Philosophy of Science* v. 1, no. 2 (Apr. 1934), 163-169.

[87] Albert Einstein, letter to Schrödinger 22 Dec. 1950. *Letters on Wave Mechanics*, Ed. K. Pribram, Trans. Martin J. Klein (New York: Philosophical Library, 1967) 39-40.

[88] Paul Arthur Schilpp, *Albert Einstein: Philosopher-Scientist* (New York: MJF Books, 1949) 75.

[89] Schilpp, *Albert Einstein: Philosopher-Scientist*, 87.

[90] quoted in Arthur Fine, "Einstein's Interpretations of the Quantum Theory," *Einstein in Context*, Eds. Mara Beller, Robert S. Cohen, and Jurgen Renn (Cambridge University Press, 1993) 257-274, 270.

[91] Max Born, *Scientific Papers Presented to Max Born on his retirement from the Tait Chair of Natural Philosophy in the University of Edinburgh* (Edinburgh, Scotland: Oliver and Boyd, 1953), 39.

[92] Albert Einstein, Trans. Jean Piccard, "Physics and Reality," *Journal of the Franklin Institute* (Mar 1936) 349-382, 377.

[93] Schilpp, *Albert Einstein: Philosopher-Scientist*, 85, 87.

[94] Einstein, *Letters on Wave Mechanics*, 56.

[95] A. Einstein, *Dialectica* v. 2, (1948) 320.

[96] Jean-Pierre Luminet, *Les trous noirs*, (Belfond, 1987) 232.

[97] The angstrom is 10^{-10} m, a ten-millionth of a millimeter. The fermi is 10^{-4} angstroms, that is, 1/10,000 of an angstrom.

[98] The parsec (abbreviated as pc) is 3.08 x 10^{18} cm, or 30 000 billion kilometers. The kpc (kiloparsec) is 1,000 pc.

[99] It is remarkable that, if Laplace had then considered objects of fixed size (instead of applying to them the same scale factor as to interdistances and velocities), he would have derived a relation of proportionality between distances and velocities, which is just the Hubble law of expansion of the Universe! Pierre-Simon Laplace, *Exposition du système du monde* (Paris: Ulan Press, 1992).

[100] This constant, $\hbar$, which has the dimension of an angular momentum, is the fundamental constant of quantum mechanics.

[101] The situation is better for composed particles, such as protons and neutrons (made of three quarks), the mass of which has been derived with a good precision from lattice quantum chromodynamics (QCD) calculations.

[102] Feynman, *QED: The Strange Theory of Light and Matter*, 152.

[103] The Compton length of a particle of mass m is $\hbar/mc$.

[104] The equation for force is written $F = ma$. The acceleration a is the ratio of a length by the square of a time (L T^{-2}), measured in m/s^2, while the dimensionality of a force is M L T^{-2}, where M represents the dimension of a mass.

[105] The relative uncertainty is on the order of 1/10,000. We currently estimate G to be about 6.674 x 10^{-11} m^3/kg s^2.

[106] This value is c = 299,792,458 m/s.

[107] A nanosecond roughly equates to 30 cm.

[108] Its numerical value, resulting from experimental measurements, is known to better than 1/1,000,000 in relative uncertainty.

[109] One calls an equation "scale invariant" if it does not change under the effect of a dilation or contraction. In such a case, the solutions often depend on scale as a power law (proportional to r^k). We will return to the concept of fractal objects, introduced by Benoit Mandelbrot in 1975, in Chapter 12.

[110] See Weisskopf.

[111] See Laurent Nottale, *Scale Relativity and Fractal Space-time* (London: Imperial College Press, 2011) Chapters V and VII; G.N. Ord, "Schrödinger's Equation and Discrete Random Walks in a Potential Field," *Annals of Physics* (August 1996).

[112] C. Auffray and L. Nottale, "Scale relativity theory and integrative systems biology: 1. Founding principles and scale laws," *Progress in Biophysics and Molecular Biology* (2008) 79-114; L. Nottale and C. Auffray, "Scale relativity theory and integrative systems biology: 2. Macroscopic quantum type Mechanics," *Progress in Biophysics and Molecular Biology* (2008) 97, 115-157; P. Turner and L. Nottale, "The physical principles underpinning self-organization in plants," *Progress in Biophysics and Molecular Biology* (2017) 123, 48-73.

[113] P. Turner and L. Nottale, "The origins of macroscopic quantum coherence in high temperature superconductivity," *Physica C* (2015) 515, 15-30.

[114] L. Nottale and T. Lehner, "Turbulence and Scale Relativity," arXiv:1807.11902. The remarkable idea to apply the scale relativity theory to turbulence, provided one works in velocity-space instead of position-space, is Louis de Montera's (arXiv:1303.3266).

[115] Perhaps what most closely approaches it is an elementary particle like the electron, such that particle physics can now analyze it. The method used consists of bombarding the electron with other particles, for example other electrons. The circle inside of which these will be scattered as a result (what we call its cross section) defines a sort of effective radius of the electron. Yet this depends on the energy used in the experiment, which is itself inversely proportional to the spatial scale of investigation. The result is amazing: whatever this scale is, one finds a radius for the electron that is about 137 times smaller (the inverse of the fine structure constant, related to the square of its electric charge).

[116] Francoise Balibar, *Einstein: la joie de la pensée* (Paris: Gallimard, 1993) 61.

[117] In fact, as p can be written in the form $p = q^{\ln p/\ln q}$, the fractal dimension is given by $D = \ln p/\ln q$.

[118] The strict definition of resolution, for example in astronomy, instead corresponds to the inverse of this radius of smoothing. In whatever case, the essential variable is the logarithm of this radius, $\log(\varepsilon)$, which simply changes its sign when ε is replaced by its inverse.

[119] The length of a fractal curve of fractal dimension D varies as a function of the resolution ε according to the formula: $L = L_0 \, (\lambda/\varepsilon)^{D-1}$.

[120] However, in the case of the quantum spacetime, while an upper scale of transition naturally appears, identified with the de Broglie scale (in accord with the classical character of macroscopic behavior), we will be led to postulate that a lower scale of transition does not exist, thus ensuring a strict nondifferentiability.

[121] This is true, as long as the fractal structure of the coast is not prolonged without discontinuity into that of the rocks: however, even in this case, a transition has every chance of existing in the form of a jump in the fractal dimension, since the rock surfaces are themselves fractal.

[122] J. Chaline, L. Nottale, and P. Grou, "L'arbre de la vie a-t-il une structure fractale?" *Comptes Rendus de l'Académie des Sciences* v. 328, Issue 11 (1999) 717-726; Laurent Nottale, Jean Chaline, and Pierre Grou, *Les arbres de l'évolution* (Paris: Hachette, 2000) 379; Laurent Nottale, Jean Chaline, and Pierre Grou, *Des fleurs pour Schrödinger : la relativité d'échelle et ses applications*, (Paris: Ellipses, 2009) 421.

[123] R. Cash, J. Chaline, L. Nottale, and P. Grou, "Développement humain et loi log-périodique," *Comptes Rendus Biologies* v. 325, Issue 5 (2002) 325, 585-590.

[124] Benoit Mandelbrot, *The Fractal Geometry of Nature* (New York: Times Book, 1982) 331.

[125] This theorem can be deduced in part from a theorem of Lebesgue (1903), according to which a curve of finite length is almost everywhere differentiable.

[126] A. Einstein, *Oeuvres choisies, v. 4, Correspondances françaises* (Paris: Seuil/CNRS) 60.

[127] A. Einstein, in *Oeuvres choisies, v. 1, Quanta* (Paris: Seuil/CNRS, 1989) 249.

[128] Richard Feynman and Albert Hibbs, *Quantum Mechanics and Path Integrals* (Mineola, NY: Dover Publications, 2005) 176-77.

[129] We will not go that far. On the contrary, one of the bases of scale relativity consists of conserving the hypothesis of continuity of spacetime, since it is this which underlies the connection between fractals and nondifferentiability (a nondifferentiable continuum is necessarily fractal), which is no longer guaranteed if the hypothesis of continuity is abandoned. But other physicists have tried to give up also continuity, in particular in the theories of quantum gravity beneath the Planck scale (Wheeler, Hawking, Rovelli, and others).

[130] One angstrom is equal to a ten-billionth of a meter. The characteristic scale of atoms is given by the Bohr radius, which is 0.529 Å.

[131] The Compton wavelength of the electron, equal to $\hbar/mc$, is 137 times smaller than the Bohr scale (the fine structure constant, square of the electric charge in adimensional units, is $1/137.0359997$).

[132] More precisely, the mass of the muon is equal to 206.77 times that of the electron.

[133] These are "internal structures" in a sense which does not put into question the unstructured, point-like character of the electron (up until energies obtained by current accelerators): these structures can be described as those of the field which surrounds them.

[134] A fermi is equivalent to a ten-thousandth of an angstrom and gives the characteristic size of an atomic nucleus.

[135] Such a number would be composed of 26 significant digits.

[136] Namely, the law of composition of velocities in special relativity (of motion) is $w = (u + v)/(1 + uv/c^2)$.

[137] Let us recall that the de Broglie length is equal to $\hbar/p$ and de Broglie time is $\hbar/E$ (which is equivalent to the Einstein relation $E = h\nu$), where p is the momentum and E the energy of the system under consideration, h being Planck's constant and $\hbar = h/2\pi$ the reduced Planck constant.

[138] But we can note that it becomes a property of the measurement precisely due to its universality, and from the fact that the measurement process is described as an interaction between the measurement apparatus, which

possesses classical *and* quantum properties, and the quantum object to be measured.

[139] For example, this limit is given by $\hbar/2$ when the resolution δx is identified to a standard error, but becomes $\hbar/\pi$ when it is identified to an interval.

[140] In other words, spacetime is just assumed to be a continuum, which may be differentiable or not. The differentiable case corresponds to classical physics. The nondifferentiable case corresponds to a fractal geometry, not in the sense of simple scale-similarity, but in the general meaning of explicit scale dependence and of metric divergence toward small scales. Therefore the fractality here does not mean a particular geometry, but on the contrary an opening toward a huge set of new geometries.

[141] Divergent functions are already well known and used in today's physics. For example, in todays's cosmology, where the universe is found to be open and without any limit, coordinates are divergent in space. The new divergence introduced in scale relativity is just the equivalent in "scale space."

[142] Even if, in certain domains of scale, this dependence can disappear (by degeneration), it is no less formally universal: this situation is analogous to the case of the static from the point of view of the laws of motion, where all explicit dependence on speed disappears, without putting into doubt the universality of the relativity of motion.

[143] As we shall see, it is natural to identify these two scales with the Planck spacetime scale (for the minimal scale) and to the cosmological constant scale (for the maximal one). Note that these scales are not limits or cut-offs but *horizons*, i.e., though being finite in their measurement, they ought to have all the properties of the zero and the infinite, being unreachable and unsurpasable.

[144] The word dilation is used here in a general sense: one assumes that the value of ρ can also be lower than 1, which would correspond to a contraction.

[145] In effect, if the length L varies proportionally to $\varepsilon^{-\delta}$, where $D = 1 + \delta$ is the fractal dimension, in a scale transformation $\varepsilon \rightarrow \varepsilon' = \rho\,\varepsilon$, the pair $(\log L, \delta)$ is transformed as $\log L' = \log L - \delta \log\rho$, $\delta' = \delta$. The Galilean transformation of motion is written $X' = X - TV$, $T' = T$. Comparing X to $\log L$ and T to δ, these

are exactly the same laws. The law of composition of dilations, $\log\rho'' = \log\rho + \log\rho'$ is also analogous to the law of composition of velocities, $V'' = V + V'$.

[146] See Michael F. Barnsley, *Fractals Everywhere* (Orlando, FL: Academic Press, 1988).

[147] The non-linear problem is that of a "general scale relativity," a theory which essentially remains yet to be constructed. The linear problem is mathematically stated in the following form. It is a matter of finding the general form of four functions $A(V)$, $B(V)$, $C(V)$, $D(V)$ where $V = \log\rho$ characterizes the change in resolution, intervening in a linear scale transformation of generalized coordinates (the position on the fractal curve and the fractal dimension $D = 1 + \delta$): $\log L' = A(V) \log L + B(V) \delta$, $\delta' = C(V) \log L + D(V) \delta$, which comes under the principle of scale relativity.

[148] Nottale, "The Theory of Scale Relativity."

[149] For a given scale of fractal/non-fractal transition λ, if one applies a dilation factor ρ to a scale $\varepsilon << \lambda$ one obtains a new scale $\varepsilon' << \lambda$ given by the formula: $\log(\lambda/\varepsilon') = [\, \log(\lambda/\varepsilon) + \log\rho \,] / [\, 1 + \log(\lambda/\varepsilon) \log\rho / \log^2(\lambda/\lambda_P) \,]$.

Whatever the scale λ of symmetry breaking, the application following this law of whatever factor of dilation ρ at the scale $\varepsilon = \lambda_P$ gives exactly the same scale λ_P, which is thus found to be invariant under dilations and contractions.

[150] Nottale, "The Theory of Scale Relativity."

[151] This problem remains unsolved in scale relativity, but it is asked in a new framework where the Planck scale *for mass* is no longer identified with the *spatiotemporal* Planck scale. Only the spacetime Planck scales become unreachable and unsurpassable limits, while the Planck energy can be crossed. This is similar to what happens in the special relativity of motion, where velocities can no longer be larger than c, while the energy can become larger than $E = mc^2$. However, physics is expected to change in a dramatic way at the Planck mass scale in the scale relativity framework as well. The curvature of spacetime (manifested as gravitation) increases toward this scale in such a way that it becomes undistinguishable from the fractality of spacetime (manifested as quantum mechanics and quantum fields). We therefore recover in this geometric way the fact that, at the Planck mass scale, quantum effects, gravitation, and quantum fields become all of the same order.

[152] This is similar to what happens in (motion) relativity: in Galilean relativity, the momentum is written $p = mv$, so that an infinite momentum corresponds to an infinite velocity. But in special relativity, the new relation is $p = mv/\sqrt{(1-v^2/c^2)}$, so that the infinite momentum is now attained when v tends to the speed of light c, while $p = mc$ no longer corresponds to $v = c$, but to the smaller value $c/\sqrt{2}$.

[153] See Laurent Nottale, *L'Univers et la Lumière* (Paris: Flammarion, 1995).

[154] In practice this is not the case, since it is linked to quantum objects which are themselves not deterministic due to its nondifferentiable character.

[155] Recall that, even in the special scale relativity framework, the spacetime Planck scale is not a limit but a horizon, i.e., there is no limit to possible successive changes of scale, and therefore to the possible structures appearing at these different scales.

[156] To be complete, we will see in the following that the particle should be considered as a twin set of geodesics corresponding to the *two* possible definitions of derivatives, which leads to a two-valuedness of the concept of velocity, and by extension, to the complex character of the wave function.

[157] Nottale, *Scale Relativity and Fractal Space-Time.*

[158] Nottale, *Scale Relativity and Fractal Space-Time.*

[159] L. Nottale, "Fractals and the Quantum Theory of Space-Time," *International Journal of Modern Physics* v. A4 (1989) 5047-5117.

[160] M.N. Célérier and L. Nottale, "Quantum-classical transition in scale relativity," *Journal of Physics A* v. 37, 3 (2004), 931-955; and M.N. Célérier and Laurent Nottale, "The Pauli equation in scale relativity," *Journal of Physics A: General Physics* v. 40 (2006) 12565-12585.

[161] L. Nottale, "Scale Relativity: First Steps toward a Field Theory," Relativity meeting, 1993, Asturias, Spain (Editions Frontieres, 1994) 121-132.

[162] L. Nottale, M.N. Célérier, and T. Lehner, "Non-Abelian gauge field theories in scale relativity," *Journal of Mathematical Physics*, 47(3) (2006) 1-19.

[163] In some texts, we have called this contribution "classical part," but this denomination can be misleading. It is classical only in the sense that it is a differentiable contribution, so we now prefer the name "differentiable part." It is only when all the effects of fractality and nondifferentiability disappear that it becomes a classical variable, since in this case it is a solution of Newton's dynamics equation of classical physics.

[164] The origin of this pairing ("Cooper pairs") is well known for standard superconductivity—it is due to the coupling of the electrons with phonons, which are the vibration modes of the lattice of atoms of the material—but remains obscure for "high" temperature superconductivity, observed for example in cuprates. We have recently suggested applying the scale relativity/fractal spacetime approach to this problem (Turner and Nottale, "The origins of macroscopic quantum coherence in high temperature superconductivity").

[165] That is to say, at the level of differential intervals which serve to describe the geodesics. One considers a given instant, then another separate from the first by an infinitesimal difference, and one describes the form taken by the elementary movements during this interval of time, as well as effects of these movements on the different physical quantities. In the scale-relativistic perspective of a nondifferentiable space, one gives the time differential element the new meaning of an explicit variable playing the theoretic role of a resolution, while in differentiable physics this element does not remain explicitly present in the equations: in standard physics, it is only an intermediary, a practical notion which disappears when one makes it tend toward zero.

[166] $V_+(t,dt) = [X(t+dt,dt)-X(t,dt)]/dt$ is *a priori* different from $V_-(t,dt) = [X(t,dt)-X(t-dt,dt)]/dt$ which is exchanged in the transformation $dt \rightarrow -dt$.

[167] Such a result should be seen with respect to Feynman's path integral, in which the probability amplitude (which is a particular form of the wave function) is calculated as an integral of a function of the classical action.

[168] The complex velocity field V is linked to the wave function ψ by the relation: $V = -2iD(grad\psi)/\psi$.

[169] Nottale, *Fractal Space-Time and Microphysics*.

[170] This operator is written $d/dt = \partial/\partial t + V.\text{grad} - iD\Delta$, where V is the complex velocity and where D is a parameter depending on the mass of the particle, $D = \hbar/2m$ (see Nottale, *Fractal Space-Time and Microphysics*).

[171] This equation is written in the form of a cancelling of acceleration $d^2x/dt^2 = 0$, where d/dt is the covariant scale derivative (see previous note).

[172] R. Hermann, "Numerical simulation of a quantum particle in a box," *Journal of Physics A* v. 30, no. 11 (1997) 3967; Nottale, *Scale Relativity and Fractal Space-Time*, Chap. 10.5: "Numerical simulation of fractal geodesics."

[173] See G.N. Ord, "Fractal Space-Time," *Journal of Physics A: General* v. 16, Number 9 (1983); Nottale, "Fractals and the Quantum Theory of Space-Time."

[174] Célérier and Nottale, "Quantum-classical transition in scale relativity." Célérier and Nottale, "The Pauli equation in scale relativity"; Nottale, Célérier, and Lehner, "Non-Abelian gauge field theories in scale relativity"; L. Nottale and M.N. Célérier, "Derivation of the postulates of quantum mechanics from the first principles of scale relativity," *Journal of Physics A: Math. Theor.* 40, (2007) 14471-14498; M.N. Célérier and L. Nottale, "Electromagnetic Klein-Gordon and Dirac equations in scale relativity," *International Journal of Modern Physics* A 25, (2010) 4239-4253; Nottale, *Scale Relativity and Fractal Space*-Time, Chaps. 5,6,7,11, and references therein.

[175] See Nottale, *Scale Relativity and Fractal Space-Time*.

[176] Nottale, *Fractal Space-Time and Microphysics*, Chapter 7.3: conclusion.

[177] See, for example, Cohen-Tannoudji and Spiro, *Matter-Space-Time*.

[178] Indeed, the unification should be made with the gravitational field as well. However, the problem of constructing a theory including quantum behavior, gauge fields, and gravitational field (a "quantum gravity") proves to be extremely difficult. Most attempts consist of trying to quantize the gravitational field, but they rely on the hypothesis that the quantum laws are still valid at the Planck scale. However, one can show that, in the same way as the quantum breaks the classical gravitation laws, gravitation breaks the quantum laws at the Planck energy-scale!

[179] In other words, why is the amplitude of the gravitational force written $F = G\, mm'/r^2$?

[180] The electric force takes the form: $F = e^2/4\pi r^2 = \alpha \hbar c/r^2$ in the case of the interaction between two electrons, with electric charge e.

[181] The gravitational force is written $F = \hbar c (m/m_p)(m'/m_p)/r^2$, where m_p is the Planck mass.

[182] Some of these grand unification bosons would keep the Planck mass at lower energy (after the symmetry breaking of the unique field), while others would not acquire mass. Among these latter particles, a fraction acquires a mass at the electroweak scale (which is 10^{17} times lower in energy than the Planck scale), forming the weak bosons, while gluons and photons remain without mass.

[183] The phase is proportional to the action, $(p\, x - E\, t + \sigma\, \varphi + e\, \chi)$, where χ is the so-called "arbitrary" gauge function, which is therefore nothing else but the conjugate of the charge e.

[184] See Nottale, *Fractal Space-Time and Microphysics*, Chap. VII-1; *Ciel et Terre: Bulletin de la Société Royale Belge d'Astronomie*, 1998; *Scale Relativity and Fractal Space-Time*, Chap. 12.

[185] For the calculation of the cosmological constant, see Nottale, *Fractal Space-Time and Microphysics*, chap. VII-1, and more recently, *Scale relativity and Fractal Space-Time*, Chap. 12. The estimate of the scale of maximal length is on the order of 3 gigaparsecs.

[186] It is written $mc^2 + (-G\, m\, \Sigma_i\, m_i/r_i) = 0$, where the sum is made on all the masses at distances in the whole universe. The inerial mass m of the body disappears from this relation, and the sum $\Sigma_i\, m_i/r_i$ becomes a ratio M_U/R_U between a typical mass M_U and a typical length R_U for the universe (which are not, strictly, its mass and its radius, since the universe is probably infinite, according to the most recent cosmological model; it may be a horizon and the mass inside this horizon). One obtains the Machian relation $(G/c^2)\, M_U/R_U = 1$, which means that the universe is in its own black hole radius.

[187] At the time of the first edition of this book. A proposal for theoretical understanding of this value has subsequently been suggested, see L. Nottale,

G. Schumacher, and E.T. Lefèvre, "Scale relativity and quantization of exoplanet orbital semimajor axes," *Astronomy and Astrophysics* v. 361 (2000) 361, 379.

[188] Nottale, *Fractal Space-Time and Microphysics*, Chap. 7.2., 333 and L. Nottale, G. Schumacher, and J. Gay, "Scale relativity and quantization of the solar system," *Astronomy and Astrophysics* v. 322 (1997) 1018-1025.

[189] R. Hermann, G. Schumacher, and R. Guyard, "Scale relativity and quantization of the solar system," *Astronomy and Astrophysics* v. 335 (1998) 281-286.

[190] Nottale, Schumacher, and Gay, "Scale relativity and quantization of the solar system."

[191] Nottale, *Scale relativity and Fractal Space-Time*, Chap. 13.

[192] Nottale, *Scale relativity and Fractal Space-Time*, Chap. 13.4.

[193] L. Nottale, "Scale relativity and quantization of extrasolar planetary systems," *Astronomy and Astrophysics* v. 315 (1996) L9-L12; L. Nottale, G. Schumacher, and E.T. Lefèvre, "Scale relativity and quantization of exoplanet orbital semimajor axes," 379-387; L. Nottale, D. Ceccolini, D. da Rocha, N. Tran-Minh, P. Galopeau, and G. Schumacher, "Structuring of the semimajor axes and eccentricities of exoplanets," *Astronomical Society of the Pacific Conference Series* v. 321 (2003) 355; L. Nottale, "Scale relativity and Fractal Space-Time: theory and applications," *Foundations of Science* v. 15, Issue 2 (2010) 101-152.

[194] Nottale, *Fractal Space-Time and Microphysics*, Chap. 7.2.

[195] Nottale, "Scale relativity and quantization of extrasolar planetary systems" and *Scale Relativity and Fractal Space-Time*, Sec. 13.5.3.

[196] Nottale "Scale relativity and quantization of extrasolar planetary systems"and *Scale relativity and Fractal Space-Time*, Chap. 13.5.3.

[197] L. Nottale, "Scale relativity and quantization of the universe. Theoretical framework," *Astronomy and Astrophysics* v. 327 (1997) 867-889.

[198] D. da Rocha and L. Nottale, "On the morphogenesis of planetary nebulae," *Revista Mexicana de Astronomía y Astrofísica* v. 15 (2003) 69.

[199] Nottale, *Scale relativity and Fractal Space-Time*, Chap. 13.7.

[200] W.G. Tifft, "Discrete States of Redshift and Galaxy Dynamics," *Astrophysical Journal* v. 206 (1976) 38-56.

[201] This may explain the apparent Tifft effect, since a probability peak in the velocity leads to a plateau for its projection, which may result in misleading apparent probability peaks due to the fluctuations for small numbers.

[202] Hermann, Schumacher, and Guyard.

[203] Nottale, Schumacher, and Lefevre.

[204] da Rocha and Nottale.

[205] Nottale, *Scale relativity and Fractal Space-Time*, Chap. 13.

[206] For a more complete account, see Nottale, *Scale Relativity and Fractal Space-Time* and references therein.

[207] F. Ben Adda and J. Cresson, *Comptes Rendus de l'Académie des Sciences* v. 330 (2000) 261; F. Ben Adda and J. Cresson, *Applied Mathematics and Computation* v. 161 (2005) 323; F. Ben Adda, *International Journal of Pure and Applied Mathematics* v. 38 (2007) 159; J. Cresson, *Journal of Mathematical Physics* v. 44 (2003) 4907; J. Cresson, *International Journal of Geometric Methods in Modern Physics* v. 3, no. 7 (2006); J. Cresson, *Journal of Mathematical Physics* v. 48 (2007) 033504.

[208] C. Alunni, *Revue de Synthèse* v. 122 (2001) 147; C. Alunni, E. Brian, and L. Nottale, "Specula," *Revue de Synthèse* v.122, 4e S., N.1 (2001) 147-183; C. Alunni and L. Nottale in "Colloquium in honor of Gilles Châtelet," *Revue de Synthèse* (2011); C. Alunni, *Advances in Applied Clifford Algebra* (2008) 18.

[209] V. Bontems and Y. Gingras, *Social Science Information* v. 46 (2007) 607-653.

[210] L. Nottale and P. Timar, "De l'objet `a l'espace psychique," *Psychanalyse et Psychose* v. 6 (2006) 195-212; L. Nottale and P. Timar, "Relativity of scales: application to an endo-perspective of temporal structures" in *Simultaneity : Temporal Structures and Observer Perspectives*, Eds. S. Vrobel, O. Rössler, T. Marks-Tarlow (Singapore: World Scientific, 2008) Chap. 14, 229-242.

[211] L. Nottale, "Generalized quantum potentials," *Journal of Physics A: Math. Theory* v. 42 (2009) 275306; L. Nottale, "Quantum-like gravity waves and vortices in a classical fluid," *arXiv* (2009) 0901.1270; L. Nottale and T. Lehner, "Numerical simulation of a macroscopic quantum-like experiment : example of the oscillating wave packet," *International Journal of Modern Physics C* v. 23, Issue 5 (2012) 1250035, 1-27.

[212] M. Forriez, P. Martin, and L. Nottale, "Lois d'échelle et transitions fractal-non fractal en géographie," *L'Espace Géographique* v. 2 (2010) 97-112; L. Nottale, P. Martin, and M. Forriez, "Analyse en relativité d'échelle du bassin versant du Gardon (Gard, France). Etude de la variation de la dimension fractale en fonction de l'altitude et de l'échelle," *Revue Internationale de Géomatique* v. 22 (2012) 103-134.

[213] L. Nottale, "Un nouveau paradigme pour la physique? Nouvelles perspectives," in *Les grands défis technologiques et scientifiques au XXIè siècle*, sous la direction de Philippe Bourgeois et Pierre Grou (Paris: Ellipses, 2007) Chapitre 9, 121-137.

[214] The application of the theory of scale relativity to biology has been greatly developed since the first publication of this book. See in particular: Auffray and Nottale, "Scale relativity theory and integrative systems biology. 1. Founding principles and scale laws," 79-114; Nottale and Auffray, "Scale relativity theory and integrative systems biology. 2. Macroscopic quantum-type mechanics," 115-157; P. Turner and L. Nottale, "The physical principles underpinning self-organization in plants," 48-73; Denis Noble, *The Music of Life* (Oxford University Press, 2008); *Dance to the Tune of Life: Biological Relativity* (Cambridge University Press, 2017).

[215] Chaline, Nottale, Grou, "L'arbre de la vie a-t-il une structure fractale?"; Nottale, Chaline, and Grou. *Les arbres de l'evolution*; L. Nottale, J. Chaline, and P. Grou, "On the fractal structure of evolutionary trees," invited conference in Fractals in Biology and Medicine, Vol III, *Proceedings of Fractal 2000 Third International Symposium, Ascona, Switzerland*, Eds. G. Losa, D. Merlini, T. Nonnenmacher, and E. Weibel, Birckhäuser Verlag (2002) 247-258; L. Nottale in "Proceedings of First International Conference on the Evolution and Development of the Universe," 8-9 October 2008, ENS, Paris, France, *Foundations of Science* v. 15 (2010) 101-152.

[216] Cash, Chaline, Nottale, and Grou, "Développement humain et loi log-périodique."

[217] Nottale, Chaline, and Grou. *Les arbres de l'evolution*; A. Johansen and D. Sornette, "Finite-time singularity in the dynamics of the world population, economic and financial indices," *Physica* A v. 294 (2001) 465-502; Nottale, Chaline, and Grou P, *Des fleurs pour Schrödinger: la relativité d'échelle et ses applications*; I. Brissaud, J. Chaline, P. Grou, and L. Nottale, "Expansion territoriales log-périodiques. Les exemples de la Russie et de la Rome antique," *Math. Sci. hum. / Mathematics and Social Sciences* v. 198 (2012) 29-48.

[218] Cf. Grant Maxwell, *The Dynamics of Transformation* (Nashville, TN: Persistent Press, 2017).

[219] D. Sornette and C.G. Sammis, *Journal of Physics I* (France) v. 5 (1995) 607; D. Sornette, *Phys. Rep.* (1998) 297, 239; Didier Sornette, *Why Stock Markets Crash* (Princeton, NJ: Princeton University Press, 2003).

[220] Nottale, *Scale relativity and Fractal Space-Time*, Chap. 14.

[221] Louis de Montera, "A theory of turbulence based on scale relativity" (2013) arXiv:1303.3266.

[222] L. Nottale, "Relativity of Scales, Fractal Space and Quantum Potentials" in *Space-Time Geometry and Quantum Events*, Ed. I. Licata (New York: Nova Publishing, 2014) Chap. 5. 175-196; T. Lehner and L. Nottale, "A new mechanics for turbulence," Présentation au Colloque Défis et aspects fondamentaux de la turbulence, IJLRA, Jussieu, Paris, 5-6 mai 2014, (org. Claude Cambon & Pierre Sagaut); L. Nottale and T. Lehner, "Intermittence lagrangienne et analogie quantique," Présentation au Workshop ERCOF-TAC/Centre Henri Bénard/SIG 35, Ecoulements turbulents, IJLRA, Jussieu, Paris, 4-5 mai 2015 (org. Claude Cambon & Thomas Gomez); L. Nottale and T. Lehner, "Turbulence and Scale Relativity," arXiv:1807.11902.

[223] Nottale, *Fractal Space-Time and Microphysics*, "Conclusion."

[224] This kind of log-periodic law applies also to many evolutive lineages of the "tree of life". See: Chaline, Nottale, Grou, "L'arbre de la vie a-t-il une structure fractale?" 717-726.

[225] In the same way as we experience three-dimensional space while our view is only two-dimensional by rotation of objects in front of us (or by rotating around them). This is clear on a computer screen, on which we can see the rotation of an object as if it were real, just by the projection effect due to rotation. This allows us to "see" the third coordinate, which remains hidden if we are at rest. In the case of spacetime, the fourth coordinate (time) would manifest itself by large 4D rotations, which correspond to speeds approaching the velocity of light. But because spacetime is not Euclidean but Minkowskian, this manifestation would be unusual, since the "projection" effect of a rotation amounts to a dilation instead of a contraction.

[226] Prigogine and Stengers; Prigogine; Le Méhauté, Nigmatullin, and Nivanen.

[227] Immanuel Kant, *Critique of Pure Reason*, Trans. J. M. D. Meiklejohn (London: Henry Bohn, 1855) 266, 265.

AFTERWORD

[228] *The Relativity Theory of Protons and Electrons* (Cambridge: Cambridge University Press, 1937) 327.

[229] Albert Einstein, "Remarks on Bertrand Russell's theory of knowledge" in *The Philosophy of Bertrand Russell*, Ed. A. Schilpp, *The Library of Living Philosophers*, La Salle, Ill. (Chicago, IL: Open Court, 1946) v. V, 279.

[230] "Reply to Criticism" in *Albert Einstein: Philosopher-Scientist*, Ed. A. Schilpp, *The Library of Living Philosophers* v. 2, La Salle, Ill. (Chicago, IL: Open Court) v. II, 683-684.

[231] *The Relativity of All Things*, xiii.

[232] "Nos petits enfants, nés dans un monde depuis longtemps relativiste, commencent à s'y mouvoir avec beaucoup plus d'aisance et de naturel que nous. [Our small children, born in a world long since relativist, begin to move within it much more easily and naturally than ourselves.]" André

Lichnerowicz, "Einstein et notre science," "Prolusione tenuta per l'inaugurazione dell'anno accademico 1979-1980 nella ceremonia solenne del 27 novembre 1979, *Accademia Nazionale dei Lincei*, "Problemi attuali di scienza e di cultur," Anno CCCLXXVI, Rome, Quaderno n. 248 (1979) 3. We will be able to ask the question of whether André Licherowicz did not show much too much optimism on this subject...

[233] On this point, and paradigmatically, see the profound and profoundly revolutionary work of Gaston Bachelard: *La Valeur inductive de la relativité* [The Inductive Value of Relativity], Paris, Vrin, 1929 (republished Paris, Vrin, "Bibliothèque des textes philosophiques," 2014, Preface by Daniel Parrochia). For an analysis, see Charles Alunni, "Relativités et puissance spectrales chez Gaston Bachelard," in *Revue de Synthèse*, Paris, Albin Michel, T. 120, no. 1 (1999) 73-110; "*La valeur inductive de la relativité* contre la *Phénoménotechnique*. L'étrange dispositif de Daniel Parocchia," in *Il senso della tecnica. Saggi su Bachelard*, Bologna, Società Edtrice Esculapio, Collana "Teoria della Cultura," Eds. Paola Donatiello, Francesco Galofaro, Gerardo Ienna (2017) 59-75; "Gaston Bachelard et Albert Einstein ou de quelques affinités sélectives," in Charles Alunni, *Spectres de Bachelard: Gaston Bachelard et l'École surrationaliste* (Paris: Hermann, "Pensée des sciences," 2018).

[234] For Langevin, see his courses at the Collège de France: *Les théories de Maxwell et de Lorentz et leurs vérifications expérimentales* (1906); *La théorie électronique des radiations et le principe de relativité* (1910); *Le principe de la relativité et les théories de la gravitation* (1915-1918 ; *Les aspects successifs et les confirmations expérimentales du principe de relativité* (1919); *Le principe de relativité et la théorie de la gravitation* (1920); *Les applications du principe de relativité aux théories de la gravitation et de l'électromagnétisme* (1921); *Physique des tenseurs* (1922). Also see his initial publications: *L'Évolution de l'espace et du temps*, Scientia (1911); *Le Temps, l'espace et la causalité dans la physique contemporaine*, Société Française de Philosophie, Séance du 19 Octobre 1911, (Paris, 1911); *Le Principe de relativité*, Étienne Chiron, Paris (1922); *L'Aspect général de la théorie de la relativité*, Conférence faite (30 Mars 1922), à l'Association Générale des Étudiants, par M. Paul Langevin, Professeur au Collège de France en présence de M. Albert Einstein (1922).

For Becquerel, see *Principe de relativité et théorie de la gravitation*, courses given in 1921 and 1922 at the École polytechnique and the Muséum national d'histoire naturelle of Paris (Paris: Gauthier-Villars, 1922); *Exposé élémentaire de la théorie d'Einstein*, Paris, Payot N°21 (1922); *Gravitation einsteinienne: champ de gravitation d'une sphère matérielle* (Paris, 1923).

For Nordmann, see *Einstein et l'univers, une lueur dans le mystère des choses* (Paris: Librairie Hachette, "Le roman de la science, " 1921).

For Lémaray, see *Le Principe de relativité*, free course given at the Faculty of Science in Marseille during the first session of 1915 (Paris: Gauthier-Villars, "Actualités scientifiques," 1916); *Leçons élémentaires sur la Gravitation, d'après la Théorie d'Einstein*, free course given at the Faculty of Science in Marseille during the fourth session of 1920 (Paris: Gauthier-Villars, 1921).

For Galbrun, see *Introduction à la théorie de la relativité. Calcul différentiel absolu et géométrie* (Paris: Gauthier-Villars, 1923). Henri Galbrun was known as an actuary and author of several works in the financial sciences.

For Bloch, see *Le principe de la relativité et la théorie d'Einstein* (Paris: Gauthier-Villars, 1922).

For Cartan, see "Sur les équations de la gravitation d'Einstein," *Journal de mathématiques pures et appliquées*, 9ème série, tome 1 (Paris: Gauthier-Villars, 1922) 141-204; "Cahiers scientifiques," *Leçons sur la Géométrie des espaces de Riemann* (Paris: Gauthier-Villars, 1928); "Actualités scientifiques et industrielles," *Le parallélisme absolu et la théorie unitaire du champ* (Paris: Hermann, 1932); "Actualités scientifiques et industrielles," *Les espaces métriques fondés sur la notion d'aire* (Paris: Hermann, 1933); "Actualités scientifiques et industrielles," *Les espaces de Finsler* (Paris: Hermann, 1934). Also see *Elie Cartan-Albert Einstein: Letters on Absolute Parallelism, 1929-1932*, Ed. Robert Debever, (Princeton: Princeton University Press – Académie Royale de Belgique, 1979).

For Darmois, see *Les équations de la gravitation einsteinienne*, Mémorial des Sc. math. fasc. 25 (Paris: Gauthier-Villars, 1927).

For Weyl, see *Die Idee der Riemannschen Fläche*, (Teubner, 1997) (zuerst 1913, in *Neuauflage mit Beiträgen* von Patterson, Hulek, Hildebrandt, Remmert, Schneider; Hrsg.: R. Remmert: TEUBNER-ARCHIV zur Mathematik, Suppl. 5, 1997); *Raum, Zeit, Materie – Vorlesungen über Allgemeine Relativitätstheorie*, 8. Auflage (Berlin: Springer 1993) (zuerst 1918,

5. Auflage 1922); *Gravitation und Elektrizität*, Sitzungsberichte (Preuss. Akademie der Wiss., Januar–Juni 1918) S. 465 (wieder abgedruckt in Lorentz, Einstein, Minkowski *Das Relativitätsprinzip*); *Was ist Materie? – Zwei Aufsätze zur Naturphilosophie* (Berlin: Springer, 1924). On Hermann Weyl and relativities, cf. Erhard Scholz (Herausgeber): *Hermann Weyl's Raum-Zeit-Materie and a general introduction to his scientific work* (Birkhäuser, 2001) (DMV Seminar Band 30); Demetrio Ria, *L'Unità fisico-matematica nel pensiero epistemologico di Hermann Weyl* (Galatina, Italy: Congedo Editore, 2005); Charles Alunni, "Hermann Weyl chez Gaston Bachelard," in *Albert Einstein et Hermann Weyl. 1955-2005. Questions épistémologiques ouvertes*, Eds. Ch. Alunni, M. Castellana, D. Ria & A. Rossi, Manduria, Barbieri Selvaggi Editori – Éditions Rue d'Um, "Collana/Collection Pensée des sciences" (2009) 13-24.

For Eddington, see "Gravitation and the Principle of Relativity," *Nature* 98 (1916) 328–330; *Report on the Relativity Theory of Gravitation* (London: Physical Society of London, 1918); *Report on the relativity theory of gravitation* (London: Fleetway press Ltd., 1920); *Space, Time and Gravitation: An Outline of the General Relativity Theory* (Cambridge: Cambridge University Press, 1920); "The Meaning of Matter and the Laws of Nature According to the Theory of Relativity," *Mind*, n.s. 29 (1920) 145–158; "The Philosophical Aspect of the Theory of Relativity," *Mind*, n.s. 29 (1920) 413-422; "A Generalization of Weyl's Theory of the Electromagnetic and Gravitational Fields," *Proceedings of the Royal Society of London*, A99 (1921) 104–122; "The Relativity of Field and Matter," *Philosophical Magazine*, 42 (1921) 800–806; "The General Theory of Relativity," *Nature*, 109 (1922) 634–636; *The Mathematical Theory of Relativity* (Cambridge: Cambridge University Press, 1923); "Can Gravitation Be Explained?" *Scientia*, 33 (1923) 315–324; "The Domain of Physical Science," in *Science, Religion and Reality*, Ed. J. Needham, (New York: Macmillan, 1925) 187–218; *Relativitätstheorie in mathematischer Behandlung* (Berlin: Springer [German transl. of 1924], 1925); "Universe: Electromagnetic-Gravitational Schemes," *Encyclopedia Britannica*, 13th ed. (London, 1926) 907–908; *The Nature of the Physical World* (New York: Macmillan, 1928); *New Pathways in Science* (Cambridge: Cambridge University Press, 1935); "On 'Relativistic Degeneracy,'" *Monthly Notices of the Royal Astronomical Society*, 95 (1935) 194–206; *The Relativity Theory of Protons and Electrons* (Cambridge: Cambridge University Press,

1936); "The Reign of Relativity: 1915–1937," Haldane Memorial Lecture (London: Birkbeck College, University of London, 1937); *The Philosophy of Physical Science*, Tarner Lectures (Cambridge: Cambridge University Press, 1938 and New York, Macmillan, 1939).

For Pauli, see "Merkurperihelbewegung und Stralenablenkung in Weyls Gravitationstheorie," *Verhandlungen der Deutschen Physikalischen Gesellschaft*, 21 (1919) 742–750; "Zur Theorie der Gravitation und der Elektrizität von Hermann Weyl," *Physikalischen Zeitschrift*, 20 (1919) 457–467; *Relativitätstheorie*, in *Enzyklopädie der mathematischen Wissenschaften*, Bd. 5, Teil 2 (Leipzig: Teubner, 1920 [English transl., *Theory of Relativity*, Dover 1956]); Review of Eddington (1924), *Die Naturwissenschaften*, 13 (1926) 273–274.

For de Donder, see *Théorie du champ électromagnétique de Maxwell-Lorentz et du champ gravifique d'Einstein* (Paris: Gauthier-Villars, 1917); *La Gravifique einsteinienne* (Ann Arbor, MI: University of Michigan Library, 1921); "Mémorial des Sciences mathématiques," fasc. 8, *Introduction à la gravifique einsteinienne* (Paris: Gauthier-Villars, 1925); "Mémorial des Sciences mathématiques," fasc. 14, *Théorie des champs gravifiques* (Paris: Gauthier-Villars, 1926); *The Mathematical Theory of Relativity* (Cambridge, MA: MIT Press, 1927); "Mémorial des Sciences mathématiques," fasc. 40, *Application de la gravifique einsteinienne* (Paris: Gauthier-Villars, 1930).

[235] P. A. M. Dirac, *Early Years of Relativity*, in *The Centennial Symposium in Jerusalem*, Eds. G. Holton & Y. Elkana (Princeton, NJ: Princeton University Press, 1982) 79-80. A full study of the philosophical context should also take into consideration the work of Hermann Weyl. For a work of research that is both technically remarkable and precise concerning this general context, I refer to the absolutely fundamental work of Thomas Ryckman, *The Reign of Relativity: Philosophy in Physics 1915-1925* (Oxford: Oxford University Press, Oxford Studies in the Philosophy of Science, 2005). For a study of the biased response of the neo-positivist school to the theories of Einstein, besides Ryckman, see Marco Giovanelli, "The Forgotten Tradition: How the Logical Empiricists Missed the Philosophical Significance of the Work of Riemann, Christoffel and Ricci," *Erkenntnis: An International Journal of Scientific Philosophy* v. 78, no. 6 (Berlin: Springer, December 2013), 1219-1257. *The*

Reign of Relativity appeared in 1921 as the title of a successful work by Viscount Haldane, published in London by John Murray.

[236] For the "unified theories," see Marie-Antoinette Tonnelat, *La Théorie du champ unifié d'Einstein et quelques-uns de ses développements* (Paris: Gauthier-Villars, 1955); *Les Théories unitaires de l'électromagnétisme et de la gravitation* (Paris: Gauthier-Villars, 1965). For more recent work, see Vladimir Vizgin, *Unified Field Theories in the first third of the 20th century* (Basel: Birkhäuser Verlag, 1994); Hubert F. M. Goenner, "On the History of Unified Field Theories," *Living Reviews in Relativity* (Berlin: Max Planck Institute for Gravitational Physics, Albert Einstein Institute Germany, 2004).

[237] Ludwik Silberstein (1872-1948) was a Polish-American physicist who helped make special relativity and general relativity staples of university coursework. His textbook *The Theory of Relativity* was published by Macmillan in 1914, with a second edition expanded to include general relativity in 1924, followed by *The Theory of General Relativity and Gravitation* (New York: Van Nostrand, 1922). At the *International Congress of Mathematicians* (ICM) in 1912 at Cambridge, Silberstein spoke on "Some applications of quaternions." Though the text was not published in the proceedings of the Congress, it did appear in the *Philosophical Magazine* of May, 1912, with the title "Quaternionic form of relativity," *Philosophical Magazine* v. 23, 790–809. The quaternions used are actually biquaternions.

[238] On this point, see Stephen G. Brush, "Why was Relativity Accepted?" *Physics in Perspective* v. 1 (Basel: Birkhäuser Verlag, 1999), 184-214.

[239] Albert Einstein, "The Method of Theoretical Physics", quoted in Gerard Holton, "Einstein, Michelson, and the 'Crucial' Experiment," *Isis* v. 60 (1969), 133-197; *Thematic Origins of Scientific Thought. Kepler to Einstein*, revised edition (Cambridge, MA: Harvard University Press, 1988) 252. For other physicists expressing similar views, cf. Hermann Weyl and Arthur Eddington among others. On these philosophical questions in physics, see Gaston Bachelard, *La Valeur inductive de la relativité* (Paris: Vrin, 1929); "Physique et métaphysique", *Septimana Spinozana* (The Hague: Martin Nijhoff, 1933); *Metafisica della matematica*, a cura di Charles Alunni e Gerardo Ienna (Roma: Castelvecchi, 2016); Charles Alunni, "Relativités et puissances spectrales," cit. in note 1; "L''École de l'ETH' dans l'œuvre de

Gaston Bachelard. Les figures spectrales d'Hermann Weyl, Wolfgang Pauli et Gustave Juvet," *Revue de Synthèse*, 5ᵉ série, année N 2 (2005) 367-389; "Gaston Bachelard face aux mathématiques," *Revue de Synthèse*, tome 136, 6ᵉ série, N° 1-2, "Philosophie et Mathématiques" (Paris: Lavoisier, 2015) 1-24; "Gaston Bachelard, ancora e ancora" in *Metafisica della matematica, op. cit.*, 19-39.

[240] "That Maxwell's electrodynamics—the way in which it is usually understood—when applied to moving bodies, leads to asymmetries which do not appear to be inherent in the phenomena is well known"; translation of Einstein's 1905 paper in Arthur Miller, *Albert Einstein's Special Theory: Emergence (1905) and Early Interpretation (1905-1911)* (Reading, MA: Addison-Wesley, 1981) ref. 8, 392.

[241] Stanley Goldberg, *Understanding Relativity: Origin and impact of a Scientific Revolution* (Boston: Birkhäuser, 1984) 189, 191; "Max Planck's Philosophy of Nature and His Elaboration of the Special Theory of Relativity," *Historical Studies in the Physical Sciences* v. 7 (1976), 125-160, cited by Brush.

[242] Stephen G. Brush, "Why was Relativity Accepted?" *op. cit.*, 193.

[243] Max von Laue, *Das Relativitätsprinzip*, (Braunschweig: F. Vieweg, 1911; second edition 1913). Later editions published under the title *Die Relativitätstheorie,* Bd. 1, *Spezielle Relativitätstheorie,* 7 Auflage (Braunschweig: F. Vieweg, 1965; 1. Auflage 1919); *Die Relativitätstheorie. Allgemeine Relativitätstheorie und Einsteins Lehre von der Schwerkraft,* Bd. 2, 5. Auflage (Braunschweig: F. Vieweg, 1965; 1. Auflage 1921).

[244] Arthur Eddington, *The theory of relativity and its influence on scientific thought* (Romanes lecture) (London: The Clarendon Press, 1922) 6.

[245] Arthur Eddington, *The Mathematical Theory of Relativity* (London, Cambridge University Press, 1923) *Preface.*

[246] Arthur Eddington, *New Pathways in Science* (London: Cambridge University Press, 1934) 211.

[247] P. A. M. Dirac, "The Early Years of Relativity," *op. cit.* 79-80; "The Excellence of Einstein's Theory of Gravitation," *Impact of Science on Society* v. 29 (1979), 11-14, 13 – cited by S. G. Brush, 202. These points of view were

equally shared and defended by Gaston Bachelard in *La Valeur inductive de la relativité, op. cit.* (1929).

[248] Quoted by Holton, *Thematic Origins, op. cit.*, 255; Abraham Païs, *'Subtle is the Lord…' The Science and the Life of Albert Einstein* (New York: Oxford University Press, 1982) 273.

[249] Arthur Eddington, *Mind*, New Series v. 29, no. 116 (Oct., 1920), 415-445, 515. Published by: Oxford University Press on behalf of the Mind Association.

[250] Arthur Eddington, *The nature of the physical world* (Cambridge, UK: Cambridge University Press, 1928) 230–231. "We have come to the point where we can say in inhabiting the plane of scientific thought renewed by relativistic hyper-criticism that *essence is a function of relation. . . . In the beginning is relation*; all realism is just a mode of expressing this relation; one can not think in a double way about the world of objects: first as relations among each other, then as existing each for themselves. . . . Relation affects being, or rather, relation is one with being." Gaston Bachelard, *La valeur inductive, op. cit.*, 208, 210, and 211. On this central core, see Nottale in the present work.

[251] Arthur Eddington, *ibid.*, 420-421.

[252] Arthur Eddington, *Mathematical theory of relativity, op. cit.*, 41. This is the so-called "structural realism" of Eddington. Eddington's epistemology is alternatively termed "structuralism" and "selective subjectivism": cf. Arthur Eddington, *The Philosophy of Physical Science, op. cit.*, VIII. For a recent discussion, cf. John Worrall, "Structural Realism: The Best of Both Worlds," *Dialectica* 43 (1989) 99-124; James Ladyman, "What Is Structural Realism?" *Studies in History and Philosophy of Science* 29 (1998) 409–424; Steven French, "Scribbling on the blank sheet: Eddington's structuralist conception of objects," *Studies in History and Philosophy of Modern Physics* 34 (2003) 227–259; Philippe Stamenkovic, "'La nature des choses' selon Eddington, ou la physique dérivée de la géométrie," Bibnum [Online], *Sciences humaines et sociales* (1 February 2012, accessed 29 March 2018. URL : http://journals.openedition.org/bibnum/875. On this point, see also Thomas Ryckman, *op. cit.*, p. 242 sq).

[253] Arthur Eddington, *The theory of relativity and its influence on scientific thought, op. cit.*, 31.

[254] Arthur Eddington, *ibid.*, 5.

[255] Arthur Eddington, *The Nature of the Physical World, op. cit.*, 122 and 123.

[256] On the connexions between Weyl and Eddington, see obviously Thomas Ryckman, *op. cit.*, passim; "Surplus Structure from the Standpoint of Transcendental Idealism: The 'World Geometries' of Weyl and Eddington," in *Perspectives on Science* (v. 11, no. 1, The Massachusetts Institute of Technology, 2003) 76-106, and Dorothy M. Wrinch, "Pure Mathematics," in *Science Progress in the Twentieth Century* (1919-1933) v. 16, no. 62 (October 1921), 173-178.

[257] On this point, see the precise analysis of Thomas Ryckman, *op. cit.*, 128 sq. Walter Becker pointed to Weyl's enigmatic designation of coordinate systems as "the unavoidable residuum of the ego's annihilation" (*das unvermeidliche Residuum der Ich-Vernichtung*). That elusive phrase, recurring in insignificantly different variants in several key texts, from 1918 through 1926, does indeed signal Weyl's broad concord with the fundamental thesis of transcendental-phenomenological idealism. Cf. *Letter to Weyl* (12 April 1923) reproduced and discussed in Paolo and Thomas Ryckman "Mathematics and Phenomenology: The Correspondence between O. Becker and H. Weyl," *Philosophia Mathematica* v. 10, 130–202. This designation gives explicit recognition to the thesis of "transcendental subjectivity," the "purified consciousness" that is the residue of the "phenomenological reduction" and the ground from which all objectivity is "constituted."

[258] Hermann Weyl, *Raum-Zeit-Materie*. 4 Auflage (Berlin: J. Springer, 1921) 8; cf. Engl. Trans. H. L. Brose as *Space-Time-Matter* (London: Methuen, 1921. Repr. ed., New York: Dover, 1953) 8.

[259] Hermann Weyl, "Philosophie der Mathematik und Naturwissenschaft," in Eds. A. Baeumler and M. Schröter, *Handbuch der Philosophie*, Abt. 2 (Munich: R. Oldenbourg, 1926) 57; cf. *Philosophy of Mathematics and Natural Science*. Rev. augmented Engl. ed., of 1926/1927 based on the trans. of O. Helmer (Princeton, NJ: Princeton University Press, 1949) 75. Cited by Thomas Ryckman, *op. cit.*, 135. On the fundamental link of this position and

Weyl's "purely infinitesimal" standpoint, cf. Ryckman, *op. cit.*, 6.3, on "Pure Infinitesimal Geometry," 149 sq, and 6.3.2, "Group-Theoretical Justification of an Infinitesimal Euclidean Metric," 154 sq.

[260] On the question of Geometrodynamics, cf. *Einstein and the Changing Worldviews of Physics*, Eds. Christoph Lehner, Jürgen Renn, Matthias Schemmel, The Center for Einstein Studies, The Einstein Studies series is published under the sponsorship of the Center for Einstein Studies, Boston University (New York: Springer, 2012); *General Relativity and John Archibald Wheeler*, Eds. Ignazio Ciufolini, Richard A. Matzner (New York: Springer, 2010); *Information and Interaction. Eddington, Wheeler, and the Limits of Knowledge*, Eds. Ian T. Durham, Dean Rickles (Switzerland: Springer International Publishing, 2017).

[261] Arthur Eddington, *The Romanes Lectures 1922. The Theory of Relativity and its influence on Scientific Thought, op. cit.*, 10-11.

[262] Arthur Eddington, *The Philosophical Aspects of the Theory of Relativity*, A Symposium by A. S. Eddington, W. D. Ross, C. D. Broad, and F. A. Lindemann, in *Mind*, New Series v. 29, no. 116 (Oct., 1920), 415-445. Published by: Oxford University Press on behalf of the Mind Association, 420.

[263] Arthur Eddington, "The meaning of matter and the laws of nature according to the theory of relativity," in *Mind* v. 29 (1920) 145–158, 153. Cited by S. French in *Studies in History and Philosophy of Modern Physics* v. 34 (2003) *op. cit.*, 230.

[264] Arthur Eddington, *The Romanes Lectures 1922. The Theory of Relativity and its influence on Scientific Thought, op. cit.*, 13.

[265] *Ibid.*, 14-15.

[266] Arthur Eddington, *Philosophy of Physical Science*, Tarner Lectures 1938, Trinity College, Cambridge (Cambridge University Press, 1939) 86-87. Fifty years later, the French neurophysiologist, who was not necessarily a great reader of Eddington, would ask the following question—Is the brain a tensor?—turning Sir Eddington "on his head," as it were, by a reversal of the problem. Here was something to reorient thought to its supposed physiological location… The neurobiological context here aims to learn how

the brain can address the problems of geometry which come about in the vestibulo-ocular reflex, where the encoding the of the rotation of the head is performed using covariant coordinates. On this point, see Alain Berthoz, "Le problème de la géométrie: le cerveau est-il un tenseur?" in *Le Sens du Mouvement* (Paris: Odile Jacob, 1997). Also see, on a technical level, the fascinating studies of A. J. Pellionisz and R. Llinàs, "Brain modelling by tensor network theory and computer simulation. The cerebellum: distributed processor for predictive coordination," *Neuroscience* v. 4 (1979) 323-248; A. J. Pellionisz and R. Llinàs, "Tensorial approach to the geometry of brain function. Cerebellar coordination via a metric tensor," *Neuroscience* v. 5 (1980) 1761-1770; A. J. Pellionisz et R. Llinàs, "Spacetime representation in the brain. The cerebellum as a predictive space-time metric tensor," *Neuroscience* v. 7 (1982) 2949-2970; A. J. Pellionisz, "Coordination: a vector-matrix description of transformations of overcomplete CNS coordinates and a tensorial solution using the Moore-Penrose generalized inverse," *Journal of Theoretical Biology* v. 101 (1984) 353-375; A. J. Pellionisz, "Neural geometry: the need of researching association of covariant and contravariant coordinates that organizes a cognitive space by relating multisensory-multimotor representation," Proceedings of the International Joint Conference on Neural Networks, Washington, DC (1989).

[267] Arthur Eddington, in *Science and Religion: A Symposium* (London: Gerald Howe, 1931) 120-121. There is a mathematical tradition of the concept of the *shadow* or *ghost*, from the *Sylvestrian Umbrae* (Shadows of Quantities) of John Sylvester up to the *Umbral Calculus* of Gian-Carlo Rota. On this use in relativity, see Gaston Bachelard, *La Valeur inductive de la relativité* (1929), *op. cit.*, 108, 112. One would have to also question the related ideas of "void" and "nothingness" (see for example Bachelard, *ibid.*, 112) as algebraic operators. On this perpetual hunt for *ghosts*, a veritable hunt for quarks, see Albert Einstein in his *Correspondance avec Michele Besso*, Letter 76 from July 28th, 1925: "The question is to find out whether this [unified] field theory is compatible or not with the existence of atoms and quanta. I have no doubt that it is true in the *macroscopic* world. If only the calculations of special cases were less troublesome! *The place is full of ghosts*, but only for the moment."

[268] Thus, with spacetime being not only curved (a gravitational field) but *fractal*, one "creates" particles and everything that exists *starting from the spacetime field itself. Letter to Lorentz,* August 14th, 1913.

[269] This is the equivalence principle: a gravitational field is locally equivalent to a field in uniform acceleration.

[270] Richard Feynman, *Quantum Mechanics and Path Integrals* (New York, Dover Publications, 1965) 175-176.

[271] Einstein, *Oeuvres choisies. 1. Quanta*, 249.

[272] Gaston Bachelard, *The Philosophy of No*, Trans. G. C. Waterston (New York: The Orion Press, 1968) 81.

[273] Gaston Bachelard, *Le Pluralisme cohérent de la chimie moderne* (Paris: Vrin, 1932).

[274] Bachelard, *The Philosophy of No*, 82.

[275] Adolphe Buhl (1878-1949), who at the age of fourteen contracted a paralysis which immobilized him for several years and cause him to walk with crutches his entire life, was in 1909 at the Toulouse faculty of science named first the chair of rational mechanics, and then, chair of differential and integral calculus which he occupied until 1945. This great autodidact submitted his first thesis in 1901, titled "Sur les équations différentielles simultanées et la forme aux dérivées partielles adjointe [On simultaneous differential equations and the associated form with partial derivatives]," and then the second, which treated "La théorie de Delaunay sur le mouvement de la lune [Delaunay's theory of the motion of the moon]" (jury: Gaston Darboux, Paul Appell, great uncle of Laurent Nottale, and Henri Poincaré). He began an editorship at the important Swiss journal *L'Enseignement des mathématiques* [The Teaching of Mathematics] starting in 1903, and assumed its direction in 1920. He was famous in particular for his work on fiber bundles, quanta, and groups.

[276] This passage is prepared by analyses to this date unsurpassed on the Heisenberg inequalities, and documented in Gaston Bachelard, *L'Expérience de l'espace dans la physique contemporaine, op. cit.* (1937).

[277] Let us note in passing that the *continuity* envisioned here is no longer to be understood as based on infinitely small points (George Cantor), but as supported by infinitely small segments (the *non-standard analysis* of Abraham Robinson). For more, see the work of Bill Lawvere and René Guitart.

[278] My emphasis.

[279] Bachelard, *The Philosophy of No*, 84-86.

[280] Bachelard, *Philosophy of No*, 87-88.

[281] On the general question of *continuity*, crucial problem for Laurent Nottale and numerous mathematicians, see the interesting work of Albert Tarantola, *The Mathematics of Continuity. From General Relativity to Classical Dynamics* (Paris: Xxxx Publications, 1995).

[282] This *simplicity* postulate is of great interest, very much present in the classical theory of relativity. For example, Eddington writes: "The frame is not the world; it is supplied by the observer and depends on him. And those *relations of simplicity*, which we seek when we try to obtain a comprehension of how the universe functions, *must lie in the events themselves* before they have been arbitrarily fitted into the frame. The most we can hope from any frame is that *it will not have distorted the simplicity which was originally present*; whilst *an ill-chosen frame may play havoc with the natural simplicity of things*. . . . We have shown that the contemplation of the world from the standpoint of a single observer *is liable to distort its simplicity*, and we have tried to obtain a juster idea by *taking into account and combining other points of view*. The more standpoints the better." In *The Romanes Lectures 1922. The Theory of Relativity and its influence on Scientific Thought*, op. cit., 12, 24.

[283] This explicit calling into question of the notion of "point" (and of "object")—replaced by the idea of the *sheaf*—should be connected to its rigorous interrogation in contemporary mathematics. Here we can consider TSR along with the mathematical theory of categories, where the "point" gives way to the *arrow* (that is, the *relation*), and the revolutionary work of the mathematician Alexandre Grothendieck. See in this regard, Charles Alunni, "Des Enjeux du mobile à l'Enchantement du virtuel – et retour," in Gilles Châtelet, *L'Enchantement du virtuel. Mathématique, physique, philosophie*, Eds. Charles Alunni, Catherine Paoletti (Paris: Éditions Rue d'Ulm, 2010), "Le

point," 31 sq. Also see, from a mathematico-philosophical perspective, the fundamental work of Fernando Zalamea, *Synthetic Philosophy of Contemporary Mathematics*, Trans. Zachary Luke Fraser (Falmouth, NY: Urbanomic, 2012).

[284] See on this point figure C8 at the beginning of *Fractal space-time and microphysics.*

[285] *Pour la science*, no. 215 (septembre 1995) 34-41.

[286] *Science et Vie*, no. 1051 (avril 2005).

[287] See the issue of *Revue de Synthèse*, "Objets d'échelle," v. 122/1 (Paris: Albin Michel, 2001), with two important articles by Laurent Nottale, "Relativité d'échelle. Structure de la théorie," 11-25 and "Relativité d'échelle et morphogénèse," 93-116.

[288] See, for example, Nottale, Chaline, Grou, *des fleurs pour Schrödinger. La relativité d'échelle et ses applications* (Paris: Ellipse Marketing, 2009); Nottale, Chaline, Grou, *Les Arbres de l'évolution* (Paris: Hachette, 2000); also see https://luth.obspm.fr/~luthier/nottale/.

BIBLIOGRAPHY

Alunni, C. *Advances in Applied Clifford Algebra* (2008) 18.

———. *Revue de Synthèse* v. 122 (2001) 147.

Alunni, C., E. Brian, and L. Nottale. "Specula." *Revue de Synthèse* v.122, 4e S., N.1 (2001) 147-183.

Alunni, C., and L. Nottale. "Colloquium in honor of Gilles Châtelet." *Revue de Synthèse* (2011).

Auffray, C., and L. Nottale. "Scale relativity theory and integrative systems biology: 1. Founding principles and scale laws." *Progress in Biophysics and Molecular Biology* (2008) 79-114.

Balibar, Francoise. *Einstein: la joie de la pensée.* Paris: Gallimard, 1993.

Barnsley, Michael F. *Fractals Everywhere.* Orlando, FL: Academic Press, 1988.

Ben Adda, F. *International Journal of Pure and Applied Mathematics* v. 38 (2007) 159.

Ben Adda, F., and J. Cresson. *Applied Mathematics and Computation* v. 161 (2005) 323.

———. *Comptes Rendus de l'Académie des Sciences* v. 330 (2000) 261

Bontems, V., and Y. Gingras. *Social Science Information* v. 46 (2007) 607-653.

Born, Max. *Scientific Papers Presented to Max Born on his retirement from the Tait Chair of Natural Philosophy in the University of Edinburgh.* Edinburgh, Scotland: Oliver and Boyd, 1953.

Brissaud, I., J. Chaline, P. Grou, and L. Nottale. "Expansion territoriales log-périodiques. Les exemples de la Russie et de la Rome antique." *Math. Sci. hum. / Mathematics and Social Sciences* v. 198 (2012) 29-48.

Cash, R., J. Chaline, L. Nottale, and P. Grou. "Développement humain et loi log-périodique." *Comptes Rendus Biologies* v. 325. Issue 5 (2002) 325, 585-590.

Célérier, M.N., and L. Nottale. "Electromagnetic Klein-Gordon and Dirac equations in scale relativity." *International Journal of Modern Physics A* 25 (2010) 4239-4253.

————. "The Pauli equation in scale relativity." *Journal of Physics A: General Physics* v. 40 (2006) 12565-12585.

————. "Quantum-classical transition in scale relativity." *Journal of Physics A.* v. 37. 3 (2004) 931-955.

Chaline, J., L. Nottale, and P. Grou. "L'arbre de la vie a-t-il une structure fractale?" *Comptes Rendus de l'Académie des Sciences* v. 328. Issue 11 (1999) 717-726.

Copernicus, Nicolaus. *On the Revolutions of the Heavenly Spheres.* Trans. A. M. Duncan. New York: Barnes and Noble, 1976.

Cresson, J. *International Journal of Geometric Methods in Modern Physics* v. 3, no. 7 (2006)

————. *Journal of Mathematical Physics* v. 44 (2003) 4907.

————. *Journal of Mathematical Physics* v. 48 (2007).

Deltete, R., and R. Guy. "Einstein's Opposition to the Quantum Theory." *American Journal of Physics* (1990) 58, 673.

Einstein, Albert. *Albert Einstein: Autobiographical Notes.* Trans. Paul Arthur Schilpp, Ed. La Salle. Chicago: Open Court, 1979.

————. *Dialectica* v. 2 (1948) 320.

————. "How I Created the Theory of Relativity." *Physics Today.* Trans. Yoshimasa A. Ono (Aug. 1982) 45-47.

————. *Letters on Wave Mechanics.* Ed. K. Pribram, Trans. Martin J. Klein. New York: Philosophical Library, 1967.

————. *Oeuvres choisies, v. 1: Quanta.* Paris: Seuil/CNRS, 1989.

———. *Oeuvres choisies, v. 4: Correspondances françaises*. Paris: Seuil/CNRS, 1989.

———. "On the Electrodynamics of moving bodies." *The Principle of Relativity: Original Papers by A. Einstein and H. Minkowski*. Trans. M. N. Saha and S. N. Bose. Calcutta UP, 1920.

———. "On the Method of Theoretical Physics." *Philosophy of Science* v. 1. no. 2 (Apr. 1934) 163-169.

———. Trans. Jean Piccard. "Physics and Reality." *Journal of the Franklin Institute* (Mar 1936) 349-382, 377.

Feynman, Richard. *QED: The Strange Theory of Light and Matter*. Princeton: Princeton University Press, 1985.

Feynman, Richard, and Albert Hibbs. *Quantum Mechanics and Path Integrals*. Mineola, NY: Dover Publications, 2005.

Fine, Arthur. "Einstein's Interpretations of the Quantum Theory." *Einstein in Context*. Eds. Mara Beller, Robert S. Cohen, and Jurgen Renn. Cambridge University Press, 1993.

Forriez, M., P. Martin, and L. Nottale. "Lois d'échelle et transitions fractal-non fractal en géographie." *L'Espace Géographique* v. 2 (2010) 97-112.

Galilei, Galileo. *Dialogue Concerning the Two Chief World Systems, Ptolemaic and Copernican*. Trans. Stillman Drake. Berkeley: University of California Press, 1967, 2nd Edition.

Hawking, Stephen. *A Brief History of Time*. New York: Bantam, 1998.

Hermann, R. "Numerical simulation of a quantum particle in a box." *Journal of Physics A* v. 30. no. 11 (1997) 3967.

Hermann, R., G. Schumacher, and R. Guyard. "Scale relativity and quantization of the solar system" *Astronomy and Astrophysics* v. 335 (1998) 281-286.

Johansen, A., and D. Sornette. "Finite-time singularity in the dynamics of the world population, economic and financial indices." *Physica A* v. 294 (2001) 465-502.

Kant, Immanuel. *Critique of Pure Reason.* Trans. J. M. D. Meiklejohn. London: Henry Bohn, 1855.

Laplace, Pierre-Simon. *Exposition du système du monde.* Paris: Ulan Press, 1992.

Lehner, T., and L. Nottale. "A new mechanics for turbulence." Présentation au Colloque Défis et aspects fondamentaux de la turbulence, IJLRA, Jussieu, Paris, 5-6 mai 2014, (org. Claude Cambon & Pierre Sagaut).

Leibniz, Gottfried Wilhelm. *The philosophical works of Leibnitz : comprising the Monadology, New system of nature, Principles of nature and of grace, Letters to Clarke, Refutation of Spinoza, and his other important philosophical opuscules, together with the Abridgment of the Theodicy and extracts from the New essays on human understanding : translated from the original Latin and French.* Trans. George Martin Duncan. New Haven: Tuttle, Morehouse, and Taylor, 1890.

Leveugle, Jules. *La Relativité, Poincaré et Einstein, Planck,* Hilbert. Paris: Editions L'Harmattan, 2004.

Lochak, G., S. Diner and D. Fargue. *L'objet Quantique.* Paris: Flammarion, Nouvelle Bibliothèque Scientifique, 1989.

Luminet, Jean-Pierre. *Les trous noirs.* Belfond, 1987.

Mandelbrot, Benoit. *The Fractal Geometry of Nature.* New York: Times Book, 1982.

Maxwell, Grant. *The Dynamics of Transformation.* Nashville, TN: Persistent Press, 2017.

de Montera, Louis. "A theory of turbulence based on scale relativity." (2013) arXiv:1303.3266.

Noble, Denis. *Dance to the Tune of Life: Biological Relativity.* Cambridge University Press, 2017.

———. *The Music of Life.* Oxford University Press, 2008.

Nottale, Laurent. *Fractal Space-Time and Microphysics.* Singapore: World Scientific 1993.

———. "Fractals and the Quantum Theory of Space-Time." *International Journal of Modern Physics* v. A4 (1989) 5047-5117.

———. "Generalized quantum potentials." *Journal of Physics A: Math. Theory* v. 42 (2009) 275306.

———. *L'Univers et la Lumière.* Paris: Flammarion, 1995.

———. in "Proceedings of First International Conference on the Evolution and Development of the Universe." 8-9 October 2008. ENS, Paris, France, *Foundations of Science* v. 15 (2010) 101-152.

———. "Quantum-like gravity waves and vortices in a classical fluid." *arXiv* (2009) 0901.1270.

———. "Relativity of Scales, Fractal Space and Quantum Potentials." *Space-Time Geometry and Quantum Events*, Ed. I. Licata. New York: Nova Publishing, 2014. Chap. 5. 175-196.

———. *Scale Relativity and Fractal Space-time.* London: Imperial College Press, 2011.

———. "Scale relativity and Fractal Space-Time: theory and applications." *Foundations of Science* v. 15. Issue 2 (2010) 101-152.

———. "Scale relativity and quantization of extrasolar planetary systems." *Astronomy and Astrophysics* v. 315 (1996) L9-L12.

———. "Scale relativity and quantization of the universe. Theoretical framework." *Astronomy and Astrophysics* v. 327 (1997) 867-889.

———. "Scale Relativity: First Steps toward a Field Theory." Relativity meeting, 1993, Asturias, Spain. Editions Frontieres (1994) 121-132.

———. "The theory of scale relativity." *International Journal of Modern Physics.* A 7 (1992) 4899-4936.

———. "Un nouveau paradigme pour la physique? Nouvelles perspectives." *Les grands défis technologiques et scientifiques au XXIè siècle,* sous la direction de Philippe Bourgeois et Pierre Grou. Paris: Ellipses, 2007. Chapitre 9, 121-137.

Nottale, L., and C. Auffray. "Scale relativity theory and integrative systems biology. 2. Macroscopic quantum-type mechanics." *Progress in Biophysics and Molecular Biology* v. 97 (2008) 115-157.

Nottale, L., D. Ceccolini, D. da Rocha, N. Tran-Minh, P. Galopeau, and G. Schumacher. "Structuring of the semimajor axes and eccentricities of exoplanets." *Astronomical Society of the Pacific Conference Series* v. 321 (2003) 355.

Nottale, L., and M.N. Célérier. "Derivation of the postulates of quantum mechanics from the first principles of scale relativity." *Journal of Physics A: Math. Theor* 40 (2007) 14471-14498.

Nottale, L., M.N. Célérier, and T. Lehner. "Non-Abelian gauge field theories in scale relativity." *Journal of Mathematical Physics* 47 (3) (2006) 1-19.

Nottale, Laurent, Jean Chaline, and Pierre Grou. *Les arbres de l'evolution.* Paris: Hachette, 2000.

———. *Des fleurs pour Schrödinger : la relativité d'échelle et ses applications.* Paris: Ellipses, 2009.

———. "On the fractal structure of evolutionary trees." invited conference in Fractals in Biology and Medicine, Vol III, *Proceedings of Fractal 2000 Third International Symposium, Ascona, Switzerland.* Eds. G. Losa, D. Merlini, T. Nonnenmacher, and E. Weibel. Birckhäuser Verlag (2002) 247-258.

Nottale, L., and T. Lehner. "Intermittence lagrangienne et analogie quantique." Présentation au Workshop ERCOF- TAC/Centre

Henri Bénard/SIG 35, Ecoulements turbulents, IJLRA, Jussieu, Paris, 4-5 mai 2015 (org. Claude Cambon & Thomas Gomez).

———. "Numerical simulation of a macroscopic quantum-like experiment : example of the oscillating wave packet." *International Journal of Modern Physics C* v. 23, Issue 5 (2012) 1250035, 1-27.

———. "Turbulence and Scale Relativity." arXiv:1807.11902.

Nottale, L., P. Martin, and M. Forriez. "Analyse en relativité d'échelle du bassin versant du Gardon (Gard, France). Etude de la variation de la dimension fractale en fonction de l'altitude et de l'échelle." *Revue Internationale de Géomatique* v. 22 (2012) 103-134.

Nottale, L., G. Schumacher, and J. Gay. "Scale relativity and quantization of the solar system." *Astronomy and Astrophysics* v. 322 (1997) 1018-1025.

Nottale, L., G. Schumacher, and E.T. Lefèvre. "Scale relativity and quantization of exoplanet orbital semimajor axes." *Astronomy and Astrophysics* v. 361 (2000) 379-387.

Nottale, L., and P. Timar. "De l'objet `a l'espace psychique." *Psychanalyse et Psychose* v. 6 (2006) 195-212.

———. "Relativity of scales: application to an endo-perspective of temporal structures." *Simultaneity : Temporal Structures and Observer Perspectives*, Eds. S. Vrobel, O. Rössler, T. Marks-Tarlow. Singapore: World Scientific, 2008. Chap. 14, 229-242.

Ord, G.N. "Fractal Space-Time." *Journal of Physics A: General* v. 16. Number 9 (1983).

———. "Schrödinger's Equation and Discrete Random Walks in a Potential Field." *Annals of Physics* (August 1996).

Pais, Abraham. *Subtle is the Lord: The Science and the Life of Albert Einstein.* New York: Oxford University Press, USA, 2005.

Poincaré, Henri. "Note." *Comptes Rendus* June 5, 1905.

———. *Science and Hypothesis.* London: Walter Scott, 1905.

————. "The Theory of Lorentz and the Principle of Reaction." *Archives néerlandaises des Sciences exactes et naturelles* ser. 2, v. 5 (1900) 252-278.

da Rocha, D., and L. Nottale. "On the morphogenesis of planetary nebulae." *Revista Mexicana de Astronomía y Astrofísica* v. 15 (2003) 69.

Schilpp, Paul Arthur. *Albert Einstein: Philosopher-Scientist*. New York: MJF Books, 1949.

Schrödinger, Erwin. "Quantisation as a Problem of Proper Values, Part II." *Collected Papers on Wave Mechanics*, Second Edition. New York: Chelsea Publishing, 1978.

Sornette, D. *Phys. Rep.* (1998) 297, 239.

————. *Why Stock Markets Crash*. Princeton, NJ: Princeton University Press, 2003.

Sornette, D., and C.G. Sammis. *Journal of Physics I* (France) v. 5 (1995) 607

Tifft, W.G. "Discrete States of Redshift and Galaxy Dynamics." *Astrophysical Journal* v. 206 (1976) 38-56.

Turner, P., and L. Nottale. "The origins of macroscopic quantum coherence in high temperature superconductivity." *Physica C* (2015) 515, 15-30.

————. "The physical principles underpinning self-organization in plants." *Progress in Biophysics and Molecular Biology* v. 123 (2017) 48-73.

Vilain, Christiane. "Huygens and Relative Motion." *Relativity in General*. Eds. J. Diaz Alonso and M. Lorente Paramo. Éditions Frontières, 1994.

Weinberg, Steven. *Dreams of a final theory: the search for the fundamental laws of nature*. London: Vintage,1993.

Weisskopf, Victor. *La révolution des quanta*. Paris: Hachette, 1989.

INDEX

Abbott, Larry, 167

Adams, John Couch, 16

Alecian, Georges, xv

Alunni, Charles, xv, 256

Ampère, André-Marie, 23

Aristotle (Aristotelian), 3-4, 8, 73, 119

Aspect, Alain, 107-8

Auffray, Charles, 258

Becquerel, Antoine Henri, 33

Bell, Alexander Graham, 24

Bell, John, 107-8, 114

Ben Adda, F., 256

Bergson, Henri, 83

Besso, Michele, 37

big bang theory, 120, 225-28

Blanchard, Philippe, 240

Bohr, Niels, 97, 110, 122

Bolyai, János, 56

Bontems, V., 256

Born, Max, 17, 92. 111-12, 207-8

Boscovich, Roger Joseph, 119

Bose, Satyendra Nath, 106

Brahe, Tycho, 4

Brissaud, Ivan, 262

de Broglie, Louis, 89-90, 92, 96, 100, 102-4, 106, 114, 125, 162, 177, 179, 200, 202, 243

Brownian motion, 17, 31-32, 148, 204

Bruno, Giordano, 5

calculus, 16-20, 168

Cartan, Élie, 230, 276

Cash, Roland, 262

Célérier, Marie-Noëlle, 209

Ceres, 246

Chaline, Jean, xv, 261

chaos, 131, 152-53, 155-56, 238, 271

Christ, 5

classical mechanics, xiv, 23, 40-41, 74, 102, 104, 108, 127, 148, 152, 164, 196, 206-7, 271

Compton, Arthur, 122-23, 170, 172, 189-90, 192, 216, 220, 222

Copenhagen interpretation, 89, 166-67

Copernicus, Nicolaus (Copernican), xiii, 3-5, 8-9, 73, 77, 79, 86

correspondence principle, 90, 99, 208

covariance, 35, 53, 57, 59, 64, 79-81, 158, 163, 168, 180, 183, 188, 206-8, 211, 261

Cresson, J., 256

de Coulomb, Charles-Augustin, 24

Curie, Marie, 33

Curie, Pierre, 33

Da Vinci, Leonardo, 265

Descartes, René (Cartesian), xiii, 4-6, 13-14, 18-20, 24, 159, 169

differentiability, 16, 20, 62-63, 112, 137, 14-42, 148, 156,

158-60, 163-68, 180, 186, 191, 195, 197, 199-208, 210-11, 266-67

Digges, Thomas, 5

Dirac, Paul, 90, 100, 105-6, 131, 162, 166, 209, 231-36

Earth, 3-4, 8-9, 12-14, 18, 22-23, 27, 32, 46, 49-52, 54, 69, 77, 79, 84, 108, 119, 150, 159, 173, 227, 241-43, 249, 261

earthquakes, 149, 261, 264-66

Eddington, Arthur, 231

Edison, Thomas, 24

Einstein, Albert (Einsteinian), xiii-xiv, 8, 14-15, 17, 21-22, 24-27, 30-32, 34-43, 46-55, 57, 59-64, 67-69, 71, 73, 77, 79-81, 83-86, 89-90, 102, 104, 106-107, 109-114, 123-26, 128-29, 135, 157-59, 161-63, 165, 168, 176, 179-81, 183, 189-90, 193, 196-97, 199, 207-8, 229-31, 234, 239, 242, 252, 268, 270

Einstein-Podolsky-Rosen (EPR) paradox, 107-8

electromagnetism, 22-25, 30, 35-37, 39, 59-61, 64, 73, 85, 106, 114, 127, 162, 170, 201, 209, 219, 221-22, 235

electroweak theory, 35, 192, 201, 209, 217-18, 220-21, 235

entanglement, 107-8, 238

equivalence principle, 27, 47-48, 50-53, 55, 57, 64, 77, 86, 123, 157, 159, 235, 239, 252,

Eris, 243-44

Euclid (Euclidean), 18, 40, 54-58, 77, 119, 148, 156-58, 189, 196, 269

Eudoxus of Cnidus, 17-18

Euler, Leonhard, 26, 265

Eveno, Eric, 258

Faraday, Michael, 23-24

de Fermat, Pierre, 18

Feynman, Richard, 93-95, 99, 122, 166-67, 169, 180, 196, 201, 209, 270

Forriez, Maxime, 256

fractal geometry, 120, 126, 128-30, 136-53, 155-60, 165-67, 169, 181, 184-86, 188-91, 193, 195-205, 210-11, 221-22, 239, 242-43, 255-58, 266

fractal dimension, 139-47, 149, 159, 167, 185-91, 196, 202, 207, 221, 231, 237, 255-57

fractal spacetime, xiv, 57, 85, 112, 128-29, 131, 147, 156-59, 180-82, 194-95, 198-200, 204, 207-11, 221-22, 229, 231-32, 237, 239, 255, 266, 271, 273

Fresnel, Augustin-Jean, 22, 125

Galilei, Galileo (Galilean), xiii, 3-4, 6, 8-11, 13-14, 16, 20, 24-25, 30-31, 36, 38-39, 42-44, 47, 49, 58-59, 73, 97, 71, 73, 76-77, 79, 83, 123, 128, 161, 175, 185-89, 196, 208, 220, 223, 230

Galopeau, Patrick, 240

Gauss, Carl Friedrich, 17, 26, 56-57, 157

Gautama, Siddhartha, 86

Gay, Jean, 240

general relativity, xiv, 14-15, 17, 21, 27, 31, 46, 48, 50, 53, 55, 57-63, 68-69, 71, 73, 80-81, 83-85, 92, 113-114, 123, 128-29, 134, 157-59, 161-63, 165, 167, 176, 193, 195, 197, 207-8, 229, 242

Gingras, Y., 256

Goldstone, Jeffrey, 235

Gordon, Walter, 100, 131, 209

Goudsmit, Samuel, 104

gravity, xiv, 3, 13-15, 17-18, 20, 23-25, 27, 46-48, 50-55, 58-64, 68-71, 73, 82-86, 102, 111, 120, 122-23, 127, 129, 131, 150, 157, 159-60, 162-63, 168, 173, 192-93, 195, 197, 203, 207-8, 218-19, 226, 229-30, 232-36, 238-39, 248-50, 254-56, 259

Grou, Pierre, xv, 261-62

Hawking, Stephen, 115, 230

Halley's comet, 16

Heisenberg, Werner, 17, 90, 92, 99-100, 103-4, 110-11, 113, 134, 167, 171, 173, 178-80, 201, 216, 228, 232-33

Heliodore, Fred, 264

Hertz, Heinrich, 24

Hibbs, Albert, 166

Higgs, Peter, 193, 235

Huygens, Christiaan, xiii, 20-21, 49, 82

Hygeia, 246

Jehle, Herbert, 240

Johansen, A., 262

Jordan, Pascual, 17, 163

Jupiter, 153-55, 243, 245-47

Kaluza, Theodor, 163

Kant, Immanuel, 5, 133-34, 272

Kepler, Johannes (Keplerian), 4, 13-15, 18, 68, 70, 239-40, 243, 249, 252, 254-55

Klein, Oscar, 100, 131, 163, 209

Koch curve, 140-42,

Kuiper belt, 243-44

Lagrange function, 206

Langevin, Paul, 90

Laplace, Pierre-Simon, 14, 119-20

Laskar, Jacques, 155

Le Verrier, Urbain, 16

Lefèvre, Eric, 240

Lehner, Thierry, 256, 266

Leibniz, Gottfried Wilhelm, 16, 18, 21, 39, 49, 83, 119, 161, 168

Lemaître, Georges, 231

Lévy-Leblond, Jean-Marc, 41, 67

Lobachevsky, Nikolai, 56

Lorentz, Hendrik (Lorentzian), 25-26, 30-31, 35-36, 38-40, 42, 47, 54, 63-64, 73, 187-93, 196, 205, 224-25, 227, 229-30. 233

Lorenz's chaotic attractor, 154

Mach, Ernst (Machian), xiii, 21, 48-50, 161, 229, 234-35

macroquantum theory, 210, 238, 242-43, 246, 255, 266, 272

Mandelbrot, Benoit, 120, 128, 137, 146, 148, 167

Mars, 154-55, 241, 243, 246

Martin, Philippe, 256

Maxwell, James Clerk, 22-25, 30-31, 37, 63

Mercury, 69, 155, 241-42

Michelson, Albert, 22-23, 32

Minkowski, Hermann (Minkowskian), 40, 55, 57

de Montera, Louis, 265

Moon, 13-14, 18, 52, 108, 120, 249

Morley, Edward, 22

Navier, Claude-Louis, 265

Nelson, Edward, 204, 240

Neptune, 16, 243

Newton, Isaac (Newtonian), xiv, 5, 13-18, 20-21, 23-24, 26, 29-30, 39, 46-52, 60, 68-71, 73, 109, 113-14, 120, 123, 128, 167-68, 178, 193, 208, 210, 255

Noether, Emmy, 7, 222

nondifferentiability (see differentiability)

nonlocality, 107-8, 131

Oersted, Hans Christian, 23

Ord, Garnet, 131, 182, 237, 271

Oresme, Nicolas, 9

Orion trapezium cluster, 251

Pais, Abraham, 31, 35

Pauli, Wolfgang, 79, 90, 105, 165

Peano curve, 138-39, 141

Planck, Max, 103-4, 121-22, 124-25, 129-32, 179, 189-90, 192-94, 210, 216-20, 223-32, 236, 239, 243, 272

Plato, 119

Pluto, 243

Pocard, Marc, 258

Poincaré, Henri, xiii, 14, 21-22, 25-36, 38-40, 49, 67, 73, 79, 119, 126, 128, 152, 168, 176, 190, 268

principle of relativity, xiii-xiv, 29-31, 33, 37-38, 42, 47, 50, 53, 64, 67-68, 71-72, 74, 78-80, 83, 85-86, 111, 128-29, 180-81, 183-84, 186-89, 211, 229, 255, 267

Pythagoras (Pythagorean), 40, 57,

quantum chromodynamics (QCD), 201, 235,

quantum gravity, 115, 192, 226, 230

quantum mechanics, xiv-xv, 17, 34, 74, 85, 89-115, 121-22, 124, 126-29, 131-32, 134, 144-45, 147, 151, 156, 159-71, 173-74, 176-80, 182, 192-97, 199-211, 216-17, 219-21, 224-26, 230-40, 256, 259, 266, 269, 271

Riemann, Bernhard (Riemannian), 17, 57-58, 60-61, 156-57, 159, 163

Rouleau, Etienne, 258

Ruelle, David, 153

Saturn, 243

scale relativity, xiv-xv, 5, 16, 19, 57, 63, 78, 112, 128-31, 133-34, 155, 161-62, 165, 168, 174, 178-79, 181, 183-84, 186, 188-90, 192-93, 195, 197, 199, 201, 203-5, 207-9, 215-23, 225-28, 230-34, 237-38, 240, 245-46, 249-52, 254-56, 260-61, 265-67, 271-72

Schrödinger, Erwin, 90-92, 100, 102, 111-12, 114, 131, 156, 162-63, 178, 204-6, 208-10, 237-39, 242, 245, 250, 252, 255, 258-61, 266

Schumacher, Gérard, xv, 240

Schwarzschild, Karl, 62

Sciama, Dennis, 234

Sornette, Didier, 149, 262, 264

spacetime, xiv, 15-17, 20-21, 25-27, 40-41, 44-45, 48-50, 55, 57-64, 73, 78, 81, 84-85, 112, 114, 124, 128-29, 133, 144, 146-148, 156-59, 161-66, 170, 173, 175, 179-82, 187, 191-201, 203-205, 207-211, 220-22, 225-26, 229, 231, 237, 239, 246, 255, 268-73

special relativity, 14, 21, 22, 26-27, 30-31, 35-37, 39-48, 53-55, 60, 73, 78, 81, 84, 108, 124, 126, 129-30, 157, 161, 176, 185, 188, 193, 209, 216, 268-69

speed of light, 14, 22-23, 27, 31, 37-38, 41-42, 44-45, 48, 70, 78, 105, 108, 121, 124, 126, 129-30, 170, 190, 193, 209, 219, 269

spin, 100, 104-8, 151, 199-201

Stokes, George, 265

Sun, 4, 13, 22, 62, 69, 124, 153, 155, 173, 242-43, 245, 249

Swift, Jonathan, 119

Takens, Floris, 153

thermodynamics, 17, 32, 72, 227, 271

Tifft, William, 253

Timar, Pierre, 256

Titius-Bode laws, 240

Turner, Philip, 258

Uhlenbeck, George, 104

Uranus, 243

Venus, 241

Voltaire, 119

Von Neumann, John, 100-1

Weinberg, Steven, 35

Weyl, Hermann, 163

Wigner, Eugene, 41

Wilson, Kenneth, 151

Wise, Mark, 167

Wolszczan, Aleksander, 248-49

Young, Thomas, 22

Zel'dovich, Yakov, 231

Zeno's paradox, 133-34

www.ingramcontent.com/pod-product-compliance
Lightning Source LLC
Chambersburg PA
CBHW070348200326
41518CB00012B/2168